The Dirty Work of Neoliberalism

The Dirty Work of Neoliberalism

Cleaners in the Global Economy

Edited by

Luis L M Aguiar and Andrew Herod

First published as volume 38, issue 3 of 'Antipode'

BLACKWELL PUBLISHING
350 Main Street, Malden, MA 02148-5020, USA
9600 Cowley Road, Oxford OX4 2DQ, UK
550 Swanston Street, Carlton, Victoria 3053, Australia

The right of Luis L M Aguiar and Andrew Herod to be identified as the Authors of the Editorial Material in this Work has been asserted in accordance with the UK Copyright, Designs, and Patents Act 1988.

First published 2006 by Blackwell Publishing Ltd

Library of Congress Cataloging-in-Publication Data has been applied for

ISBN-10: 1-4051-5636-8
ISBN-13: 978-1-4051-5636-3

A catalogue record for this title is available from the British Library.

Set in 11/13pt times
by Graphicraft Limited, Hong Kong

The publisher's policy is to use permanent paper from mills that operate a sustainable forestry policy, and which has been manufactured from pulp processed using acid-free and elementary chlorine-free practices. Furthermore, the publisher ensures that the text paper and cover board used have met acceptable environmental accreditation standards.

For further information on
Blackwell Publishing, visit our website:
www.blackwellpublishing.com

Contents

Section 3

Introduction:
Cleaners and the Dirty Work of Neoliberalism

Andrew Herod and Luis L M Aguiar

Recently, Hollywood has glamorized the life of cleaners in the form of the romantic comedy *Maid in Manhattan*, starring Jennifer "JLo" Lopez and Ralph Fiennes. Little more than a rehashed Cinderella story, the movie concerns the fortunes of Marisa Ventura (Lopez), "a struggling single mom who works at a posh Manhattan hotel and dreams of a better life for her and her young son" and who, "one fateful day", bumps into hotel guest and senatorial candidate Christopher Marshall (Fiennes), who mistakes her for a wealthy socialite. As the advertising blurb puts it:

> After an enchanting evening together, the two fall madly in love. But when Marisa's true identity is revealed, issues of class and social status threaten to separate them. Can two people from very different worlds overcome their differences and live happily ever after? (Sony Pictures 2002).

Of course, this is not the first time Hollywood has chosen to focus upon the professional cleaning industry (Aguiar 2005). Thus, *Good Will Hunting* revolves around a janitor who is also a mathematical genius, one who solves complicated formulae when not cleaning the halls of the Massachusetts Institute of Technology. Significantly, though, despite the glamorization of the life of cleaners in these films—in *Maid in Manhattan* cleaners zing *bons mots* at each other as they work, whilst in *Good Will Hunting* Will (Matt Damon) gets to date a medical student played by the effervescent Minnie Driver—in both, the main premise is, actually, the possibility of escaping the janitorial life for bigger and better things.

Whilst Hollywood has often chosen to glamorize cleaners (though Ken Loach's *Bread & Roses* is an exception), for its part the cleaning industry has drawn inspiration from Hollywood. Hence, in a case of life imitating art, Kimberly-Clark, one of the largest commercial cleaning companies globally, proudly touts its "Golden Service Awards" as "The Oscars of the Cleaning Industry". Initiated in 1991, such awards "are acknowledged internationally as the hallmark of quality and a powerful tool in the promotion of excellence", with companies from the UK, Ireland, Australia, New Zealand, Belgium, Holland, and, most recently, Italy able to enter in categories such as "Educational Establishment", "Healthcare", "Manufacturing", "Transport", and others. One lucky cleaner even gets to be designated "The Kimberly-Clark Professional Cleaner of the Year" (Kimberly-Clark 2005).

Yet, much as the life of a street prostitute is undoubtedly not quite as glamorous as Julia Roberts seemed to suggest in the film *Pretty Woman*, movies such as *Maid in Manhattan* and *Good Will Hunting* present a decidedly unrealistic view of professional cleaning work, particularly in the age of neoliberalism. Thus, whereas cleaning is often thought to be easier work than "heavier" activities such as manufacturing, in actuality it is one of the most injury-prone occupations in the contemporary labor market. In the United States, for instance, in 2003 janitors and cleaners suffered some 127,800 musculoskeletal disorders involving days away from work, the fourth-highest rate of all US occupations, whereas maids and housekeepers had the tenth highest rate (US BLS 2005a, Table B).[1] Overall, janitors and cleaners had the fifth-highest number of injuries and illnesses with days away from work, after laborers and material movers, heavy and tractor-trailer truck drivers, nursing aides, orderlies, and attendants, and construction laborers (Wiatrowski 2005, Chart 1). In return, they found themselves in the bottom quartile of all US wage earners (Hecker 2005, Table 3).[2] Moreover, despite the fact that the janitorial and cleaning sector is predicted to be one of the ten fastest growing sectors for employment over the next decade (US BLS 2005b, Table 3d), the prospects for the vast majority of cleaners to improve their situation are decidedly slim, as subcontracting, intensification of labor, and attacks on unions mean such workers' jobs are increasingly taking on the characteristics of sweatshop work.

Significantly, such phenomena are not confined to the US. In Australia in 2000–2001, for instance, cleaners had an incidence of work-related injury that was 2.2 times the national average (Commonwealth of Australia 2005), whilst in the United Kingdom cleaners are some of the lowest paid of all workers, with full-time male and female cleaners paid an average of £272 and £222 per week in 2003, compared with the national averages for all occupations of £525 and £396 respectively, and with 75% of full-time male and 90.2% of full-time female cleaners earning less than £310 per week (compared with 24.1% and 43.0% of all occupations) (UK Office of National Statistics 2003, Tables D1, D2, D7, and D8). Despite such low pay and harsh conditions, though, cleaners are situated at an important nexus of the global economy, for they are essential to ensuring that the spaces of production, consumption, and social reproduction which define the social architecture of the contemporary global economy remain sanitary and functional. Indeed, without their labor, the offices, factories, hospitals, shopping malls, sporting arenas, and other spaces of the modern global economy would soon become buried under a surfeit of paper, metal shavings, medical waste, plastic packaging, and other assorted detritus of economic activity. Given their importance for the operation of the global economy, then, it is surprising that, for many, cleaners remain largely invisible in the landscape—most of us know when somewhere has not been cleaned but few of us, we suspect, stop to think much about the laboring processes which go into maintaining spaces as clean. In light of this, through this collection we hope both to draw attention to the important work that cleaners do and to examine how neoliberalism is transforming this work and worsening their economic position and, frequently, their health.

Neoliberalism and the Destruction of Fordist Work Organization

In the political-economic realm, the past two decades or so have been dominated by two words: neoliberalism and globalization. Frequently, these two words have been conflated, though in reality they refer to quite different things. Although there are a number of elements contained under the rubric of "neoliberalism" and different commentators use the term to refer to slightly different things, generally there are five characteristics which are central to the pursuit of a neoliberal agenda, these being: belief in the power of the market to most efficiently allocate resources and to encourage economic development (defined as the establishment of a market-based economy rooted in the defense of private property rights); the privatization of state-owned enterprises in the belief that this will encourage market forces and so stimulate economic efficiency; "deregulating" (which often, actually, simply means regulating in a different way) the economy—particularly labor markets—so as to limit the "distorting" effects of governmental intervention; the cutting of state expenditures on social welfare provisions; and an ideological attack upon notions of collectivism and an ideological support for the values of economic individualism—perhaps most famously expressed in Margaret Thatcher's widely reported comment that there "is no such thing as society". Such elements have been central to the "Washington Consensus" vision of globalization.

Certainly, these elements play out differently in different places, as the particularities of place mean that neoliberalism is not a spatially uniform project (Castree 2005). Nevertheless, a key aspect of the neoliberalization of labor markets globally has been the exteriorization of economic relationships which were previously internal to the structure of companies. Thus, whereas during the heyday of Fordism many firms were highly vertically-integrated and the service activities which aided them in the pursuit of their primary activities (such as cleaning or staffing workplace cafeterias) were typically conducted by workers employed by that firm (Wial 1993), today such activities are increasingly being subcontracted out and/or treated as quasi-independent (with the result that workers and managers in these areas must compete with outside contractors for continued work), all with the goal of reducing labor costs both directly in terms of wages but also indirectly in terms of pension obligations and the outlay for other benefits. The result of such a transformation in the structure of firm organization has been an immiseration of workers and, usually, an intensification of their work, one which has increased the likelihood of on-the-job injuries as workers are often forced to rush in order to complete their tasks. Significantly, this latter phenomenon has been exacerbated by two other aspects of neoliberalism.

First, the dismantling of many government regulations covering the workplace has undermined many of the protections which workers had come to rely upon as part of their "industrial citizenship" rights. This has been achieved either by lowering standards or by making formerly compulsory standards voluntary, in the belief that firms will find it in their best interests to protect their workers or risk being punished by "the market"—this latter attitude, for instance, has been the current Bush administration's approach to ergonomic standards. Second, the neoliberal assault on collectivism—carried out both by direct changes to labor laws to make it more difficult for unions to organize and/or easier for them to be decertified (often under the guise of giving workers greater "choice"), and through the privatization of public sector activities, a process which almost invariably reduces rates of unionization—has undermined labor unionism. The result has been a precipitous decline in union density which, in turn, has helped increase the rate of workplace injury—in Britain, for instance, rates of injuries for cleaners in unionized workplaces are about half those of non-unionized workplaces (5.3 injuries per 1000 employees compared with 10.9 injuries; Reilly, Paci and Holl 1995).

Professionalizing Cleaning Work

Whilst the intensification of employment and reductions in wages are transforming cleaning work, such work is also being impacted in other ways. In particular, whereas cleaning has long been taken to be fairly unskilled work there has been a noticeable move recently to "professionalize" the industry. Several factors seem to be at play here —the effort of subcontractors to give themselves purchase in the market (after all, who wants to hire an "unprofessional" outfit?), perhaps a search for authenticity within a post-modern world (ensuring that a client gets a "real" cleaning company rather than a "fly-by-night" contractor), the growing transnationalization of cleaning capital and the concomitant push for international standards, and the need for cleaners to be trained to operate new equipment being introduced as part of the intensification of cleaning work. Indeed, it is fair to say that what may once have been considered to be the art of cleaning is increasingly being viewed by many as a science. For instance, cleaners in the UK can now secure training at the British Institute of Cleaning Science on the campus of the University of York (where a former laundry has been converted into the "Cleaning Science Training School"), with trainees able to take classes on "chemical competence", "safe use and care of machines", and "suction cleaning and single solution mopping" (BICSc 2005), whilst for their part the Building Service Contractors of New Zealand have established a Certificate in Cleaning and Caretaking—emphasizing issues of health and safety, legal requirements, and how to utilize the new "cleaning products and techniques that are required to meet the needs of modern building surfaces", the Certificate is designed "to lift the status of cleaners through a nationally recognised qualification" (Young 2004).

The growth of a discourse concerning professionalization within the context of a significant neoliberalization of working conditions raises interesting questions, particularly in an environment in which neoliberal globalization is typically presented as encouraging the deregulation of work environments. Thus, whereas government "deregulation" of labor markets has facilitated the growth of outsourcing, "flexibility", and reduced wages for cleaners, the pressures to professionalize the industry are leading to a different type of regulation. Indeed, as one industry insider—whose company has won several cleaning awards, including one for maintaining the Globe Theatre in London— has stated, the industry "is becoming a regulation led profession" (Cooke 2005). Part of the reason for this appears to be a contradiction in the processes of globalization and neoliberal deregulation, specifically the fact that whereas advocates of neoliberal globalization often present the process as one in which regulations are eroded so that "the market" may allow the economically strong to emerge victorious, in actuality such anarchy can pose significant problems for transnational capital. Indeed, as much as its advocates like to portray neoliberal globalization as being about sweeping away regulations, it is clear that the process actually relies upon the establishment of international standards by which products produced in one part of the world can interface with those from another. Hence, as transnational corporations have expanded their investments into myriad nations across the planet, there has been concern on the part of many such advocates that perhaps the Pandora's Box of neoliberal globalization has been opened just a little too wide. Consequently, some have argued that it is necessary to have the "protection and security of a rules-based system" in the international arena lest corporations fall victim to the "rough and tumble of a global jungle where only the powerful survive" (Weekes 1996).

In the case of many products, such a system of rules is provided by the International Organization for Standards (ISO), which is based in Geneva, Switzerland, but which comprises national standards institutes from some 130 countries.[3] For the most part, the majority of the ISO's standards—which are voluntary, though widely followed—are technical and relate to the specifications various products should meet (Neumayer and Perkins 2005). However, its 9000 series standards relate to procedures for conducting certain management practices, and cleaning companies have increasingly sought certification as ISO 9000-compliant as a way to present an aura of professionalism to prospective clients. Indeed, ISO certification is becoming the *sine qua non* for cleaning companies wishing to distinguish themselves within the throng of firms—many of them small operations— that have emerged thanks to deregulation of the industry: whereas ISO certification is taken to be the mark of professionalism, its Other— noncertification—is viewed as marking a lack thereof. "Professionalism", in other words, has increasingly become an important marketing tool and means of corporate branding as the number of firms in the industry has exploded in recent years.

If the adoption of ISO standards represents one way in which a discourse of professionalism is permeating the industry, the transnationalization of cleaning capital is another. Hence, as companies like the Australian Spotless Services and the Danish firm ISS (which presently operates in 44 countries, its latest expansion being into India) have extended their operations globally, they have brought with them various sets of cleaning practices which have often become *the* standard ways of operating in particular geographical regions or segments of the industry, such that to be considered "professionals" local companies must match their standards—a point noted by ISS's Vice President for Human Resources, who has argued that "ISS is often competing with companies on a local level, and one of our main advantages is the training we can offer to our employees. We have in many of the countries that we are operating in been the pioneers in developing good training systems for both frontline employees and managers [and] are keen to make this one of our trademarks" (Christensen nd). In the case of such global cleaning giants' entrance into the cleaning market in developing countries, this discourse of professionalism has been facilitated by long-standing and deeply engrained associations between foreign ways of doing things and higher quality, and has encouraged a growing standardization of ways of working—in Chile, for instance, one of ISS's marketing strategies has been to laud its professionalism and to argue that its "Scandinavian values" of honesty, initiative, responsibility, and high-quality services make its product unique and something to which other companies (both foreign and Chilean) should aspire. Interestingly, and somewhat ironically, however, this top-down imposition of a professional ethics of work often actually leaves the workplace less clean than it might otherwise be, since it dismisses cleaners' own definitions of what constitutes "cleanliness" and undermines their sense of pride in performing good work (Hood 1988).

The apparent professionalization of cleaning has important implications for how cleaners are viewed, particularly concerning whether their work is seen as skilled for, as Gaskell (1986) has argued, "skill" should not be seen as some kind of "independent variable" which is a fixed attribute of a particular job but as a politically contested designation that reflects who is doing the work. Thus, as Coyle (1985:7) shows, men who work as cleaners are often not called such but are employed instead as "general maintenance workers", which allows them to secure different gradings and better pay relative to female "cleaners". In the case of cleaners, then, there appears to be an interesting contradiction emerging. On the one hand, the fact that the cleaning labor force is overwhelmingly made up of women and non-white (often immigrant) workers—increasingly so in cities which are receivers of migrants from the global South—encourages it to continue to be defined as an unskilled job. On the other hand, the push towards professionalization and the introduction of new technologies to facilitate work intensification is not only changing the physical requirements of the work but is also changing the discursive context within which cleaning is done—one illustrated by the fact that increasing numbers of operators now refer to their cleaning crews as "sanitation engineers" (Stier 2004).

Resisting Neoliberalism

As in many industries, the number of cleaners who are union members has been declining in most industrialized nations over the past 15 years or so, the result of anti-union activities by employers and the growth in the number of small, family-owned firms which are typically non-union operations. Nevertheless, it is important to recognize that there have been several notable organizing campaigns in recent years involving cleaners, perhaps the most famous of which is the Justice for Janitors (JfJ) campaign in the United States, a campaign which has

not only served to challenge traditional models of union organizing—
JfJ organizers, for instance, seek to organize according to geographic
area rather than on a workplace-by-workplace basis—but which has
also increasingly been adopted overseas. Indeed, in one of the international labor movement's more innovative moves of late, organizers
and researchers have actually been dispatched by the JfJ campaign to
work directly with cleaning unions in New Zealand and Australia,
with the goal of creating an organizational structure which matches
that of employers.

As many cleaning companies and the office firms with which they
contract have become global companies, similar moves towards
greater international labor collaboration have been embarked upon
elsewhere. Thus, cleaners' unions in the US and Canada have
attempted to develop stronger links as US chains have purchased or
built Canadian hotels, whilst within the European Union the establishment by companies such as ISS of European Works Councils is
laying the basis for possible future Europeanwide bargaining. Part of
the impetus for such transnationalization of bargaining has come
from international union organizations such as the Union Network
International (UNI), a global federation of unions representing some
15 million members worldwide in more than 900 unions in 150
countries which is working through its Property Services division to
encourage unionization amongst cleaners. Hence, in 2003, as part of
an agreement with UNI, ISS committed itself to observing 12 fundamental workplace rights, including support for minimum wages,
respect for the right to join and form trade unions, and avoidance of
excessive working hours, whilst UNI agreed, in return, to support
publicly companies which recognize such rights and to condemn
publicly those which undermine them (ISS 2003).

Certainly, then, the international labor movement is making efforts
to counter the effects of neoliberalism. However, the transformed
context within which cleaners' unions must operate relative to even a
few years ago is making organizing much more difficult and forcing
unions to devise new strategies. Given that the practices of neoliberalism appear to be deepening, it is likely that the future will pose
many challenges to cleaners and their unions as they struggle to
improve their lives and working conditions. As they engage in such
efforts, though, perhaps they will bear in mind the words of the
nineteenth century abolitionist Frederick Douglass: "If there is no
struggle, there is no progress".

Acknowledgments

The editors would like to thank all of the authors for contributing to
this collection, together with the reviewers who provided feedback on

these papers. We would also like to thank Noel Castree, *Antipode* editor, for his support of this project. We would particularly like to express our gratitude to Ros Whitehead for helping shepherd this project to completion. This was the last project for *Antipode* upon which Ros worked, and she stayed on after relinquishing her other duties just to make sure that the collection saw the light of day. For that we are both extremely grateful. Finally, we would like to thank our families for living with "cleaners" for two years: in the case of Luis—Alice, Gabriella, Juliano, and Emilio; and in the case of Andy—Jennifer.

Endnotes

[1] The Bureau of Labor Statistics makes a distinction between "janitors and cleaners" and "maids and housekeeping cleaners" (US BLS nd): the former "perform a variety of heavy cleaning duties, such as cleaning floors, shampooing rugs, washing walls and glass, and removing rubbish [and] may fix leaky faucets, empty trash cans, do painting and carpentry, replenish bathroom supplies, mow lawns, and see that heating and air-conditioning equipment works properly", whereas the latter "perform any combination of light cleaning duties to maintain private households or commercial establishments, such as hotels, restaurants, and hospitals, clean and orderly". In this collection we use the term to refer generally to workers who engage in cleaning activities as part of a job outside their own homes.

[2] In 2002, median annual earnings of janitors and cleaners were US$18,250 compared with US$16,440 for maids and housekeepers, these latter predominantly being female (US BLS nd).

[3] ISO (nd) states that its "purpose is to facilitate international trade by providing a single set of standards that people everywhere would recognize and respect".

References

Aguiar L L M (2005) Cleaners and pop culture representation. *Just Labour: A Canadian Journal of Work and Society* 5(Winter):65–79

BICSc (British Institute of Cleaning Science) (2005) *The British Institute of Cleaning Science 3M Award for the Pursuit of Excellence through COPC Training.* http://www.york.ac.uk/admin/tdo/general/3M%20Award.pdf (last accessed 16 December 2005)

Castree N (2005) The epistemology of particulars: Human geography, case studies and "context". *Geoforum* 36(5):541–544

Christensen M (nd) Statement by Martin Christensen, ISS Vice President Human Resources. http://www.issworld.com/view.asp?ID=1596&mID=1579 (last accessed 30 December 2005)

Commonwealth of Australia (2005) Occupation by occurrence result (number, incidence and frequency), new workers' compensation cases reported, Australia 2000–01. Department of Employment and Workplace Relations, Office of the Australian Safety and Compensation Council: Canberra ACT. http://www.nohsc.gov.au/ohsin formation/databases/compsuptables/2000-01/Occurrence_Result/aus_occupation_ fnf.pdf (last accessed 10 December 2005)

Cooke D (2005) Statement by Douglas Cooke, Managing Director, Principle Cleaning Services, London. http://www.kcprofessional.com/uk/?PageRequest=about_us/ goldenservice.htm (last accessed 16 December 2005)

Coyle A (1985) GOING PRIVATE: The implications of privatization for women's work. *Feminist Review* 21:5–23

Gaskell J (1986) Conceptions of skill and the work of women: Some historical and political issues. In R Hamilton and M Barrett (eds) *The Politics of Diversity* (pp 361–380). London: Verso

Hecker D E (2005) Occupational employment projections to 2014. *Monthly Labor Review* 128(11/November):70–101

Hood J (1988) From night to day: Timing and the management of custodial work. *Journal of Contemporary Ethnography* 17:96–116

ISO (International Organization for Standardization) (nd) ISO 9000: An introduction. http://www.praxiom.com/iso-intro.htm (last accessed 31 December 2005)

ISS (2003) ISS and UNI in global deal to raise industry standards. Press release, 9 May. http://www.issworld.com/view.asp?ID=1600&mID=1579 (last accessed 22 January 2006)

Kimberly-Clark (2005) The Kimberly-Clark Professional Golden Service Awards— The Oscars of the cleaning industry. http://www.kcprofessional.com/uk/ ?PageRequest=about_us/goldenservice.htm (last accessed 16 December 2005)

Neumayer E and Perkins R (2005) Uneven geographies of organizational practice: Explaining the cross-national transfer and diffusion of ISO 9000. *Economic Geography* 81(3):237–259

Reilly B, Paci P and Holl P (1995) Unions, safety committees and workplace injuries. *British Journal of Industrial Relations* 33(2):275–288

Sony Pictures (2002) *Maid in Manhattan*. http://www.sonypictures.com/homevideo/ maidinmanhattan/title-navigation-2.html (last accessed 6 December 2005)

Stier R F (2004) Clean operation: Cleanliness is next to Godliness and essential to assure safe food. *Food Safety Magazine* April/May. http://www.foodsafetymagazine. com/issues/0404/featcol0404.htm (last accessed 20 January 2006)

UK Office of National Statistics (2003) *Labour Market New Earnings Survey 2003— Data for 2003, Analyses by Occupation*. http://www.statistics.gov.uk/downloads/ theme_labour/NES2003_Analyses_By_Occupation/NES2003_Analyses_By_ Occupation.pdf (last accessed 12 December 2005)

US BLS (United States Bureau of Labor Statistics) (nd) Building cleaning workers. http://www.bls.gov/oco/pdf/ocos174.pdf (last accessed 8 December 2005)

US BLS (United States Bureau of Labor Statistics) (2005a) Lost-worktime injuries and illnesses: Characteristics and resulting time away from work, 2003. http:// www.bls.gov/news.release/osh2.nr0.htm (last accessed 8 December 2005)

US BLS (United States Bureau of Labor Statistics) (2005b) The 10 occupations with the largest job growth, 2004–14. http://www.bls.gov/news.release/ecopro.t06.htm (last accessed 8 December 2005)

Weekes J (1996) Statement by John Weekes, Canadian Permanent Representative to the World Trade Organization and the Office on the United Nations in Geneva, to the annual foreign policy conference of the Canadian Institute of International Affairs, Hamilton, Ontario, 26 October

Wial H (1993) The emerging organizational structure of unionism and low wage services. *Rutgers Law Review* 45(Summer):671–738

Wiatrowski W J (2005) Occupational safety and health statistics: New data for a new century. *Monthly Labor Review* 128(10/October):3–10

Young B (2004) Statement by Brian Young, Building Service Contractors of New Zealand. http://www.kiwicareers.govt.nz/jobs/14a_pcu/j64521e.htm (last accessed 16 December 2005)

Section 1

Chapter 1

Introduction: ·
Geographies of Neoliberalism

Andrew Herod and Luis L M Aguiar

Although as a political and economic project neoliberalism operates across global space, this is not, of course, the only scale at which its practices are manifested and felt, for whilst there are many constants to neoliberalism across the planet (such as how its ascendance is linked to working-class political defeats; Carroll 2005), there are also important differences in how neoliberalism is operationalized in different places (Brenner and Theodore 2002; Castree 2005). Put another way, neoliberalism is a spatial project that is spatially projected because, despite the rhetoric of how neoliberal globalization is purportedly producing a flat and borderless world (Friedman 2005; Ohmae 1990) in which distance and geography no longer matter, the sway of place still shapes how political praxis is imagined and articulated in these neoliberal times—the histories of social struggles and their institutional memories are very much tied up in the spatialities of the global economy and greatly influence how neoliberalism is being implemented locally and nationally (Brenner and Theodore in Keil 2002:582).

Given that this is the case, in this first section we present four papers, each of which spatializes the content and form of neoliberalism and the consequences of this for cleaners within four quite different national contexts to show how hard-won gains in the areas of social and industrial citizenship are being dismantled by an aggressive state bent on institutionalizing a "paradigm shift" towards a neoliberal "national policy" in each country (McBride 2005), one which spawns inequality through economic deregulation in the name of promoting entrepreneurialism and worker "freedom". The result of such policies is very much a return to a 19th-century *laissez-faire* capitalism of rapidly growing economic polarization and a repressive state dedicated to a minimalist social regulatory regime in which citizens are redefined as consumers and in which the economic and social risks of employment are more and more assumed by the individual worker, who is increasingly treated as an independent contractor responsible for his/her own healthcare and pension than an employee for whom an employer has some social or economic responsibility. As the papers in this section show in their different ways, then, the role of the state in all this is to facilitate the greater accumulation of capital by removing obstacles that may hinder this process, be these obstacles socialized (as in socially erected labor market shelters) or individualized (as in an individual's "poor" work ethic).

As the papers in this section illustrate, cleaning companies are taking full advantage of the deregulation of labor markets and the contracting out features of neoliberalism both to "professionalize" the industry with new and improved technologies for the changing cleaning labor process (Aguiar 2001)—such as fitting cleaners with new crisp, clean, and bright uniforms—and to slough off the responsibilities which the Fordist social contract had previously required of them. This presents an interesting paradox, for whereas one would have thought that their new professionalized attire and equipment would make cleaners more visible, the industry actually instructs them on how to make themselves imperceptible in the work environment so as not to disrupt the shopping and other fantasies of the middle classes whilst it simultaneously dissolves its social obligations to them. Each of the four papers, then, in different ways explores how neoliberal practices—specifically with regard to the (de)regulation of cleaning labor markets—are impacting the industry and its workers.

In the first paper, Luis Aguiar argues that neoliberalism represents the most recent in a long line of national socio-economic plans for Canada, one that has been implemented during the past two decades or so irrespective of the political party in power. In the case of the cleaning industry, he shows how a neoliberalized state is entrenching cleaning work as sweatshop work through deregulating labor markets, undermining workplace protections, and purposely creating a casual labor force. Through an analysis of labor market deregulation in Ontario and British Columbia, he illustrates how many of the social and industrial rights long available to cleaners under the aegis of the Fordist welfare state are being eroded. The result, he suggests, is that cleaners (and other workers) are increasingly having to endure a form of "sweatshop citizenship" in which they are left with little leverage to make demands for better working conditions, protections, and a living wage from capital and the state. This raises important questions about the nature of democracy within and without the workplace, questions which put paid to the neoliberal rhetoric that labor market deregulation is liberating workers from the constrictions of the "totalitarianism" of the welfare state by presenting them with greater freedom of choice to pursue their own economic and political destinies.

The neoliberal features of cleaning work described by Aguiar in the Canadian case, are, unfortunately, also common in the case of cleaners in South Africa, as discussed by Andries Bezuidenhout and Khayaat Fakier. In exploring the putatively post-apartheid situation, the authors make two significant arguments. First, whereas there was hope when formal apartheid ended in 1994 that the post-apartheid era would end the kinds of racial segmentation of labor markets that had dominated throughout the twentieth century, this has not actually taken place. Rather, as neoliberalism has penetrated the economy there has been a reinscription of the black body as "properly" fitted for specific industrial sectors and work locations, such that the differences between how labor markets operated under the apartheid regime and how they do so today are often minimal. Second, the restructuring of work has displaced "risk" from capital and the state onto the individual worker and his/her household management. This has meant that women particularly must intensify their own domestic work by squeezing more out of the paltry wages they earn (such as preparing from scratch more of their food, which is more time-consuming than using more expensive pre-prepared provisions), whilst also inventing other sources of income to supplement their household revenues which have been reduced through growing casualization and insecurity in the labor market. As Bezuidenhout and Fakier show, then, neoliberal policies are undermining efforts to deracialize the workplace as subcontracting reintroduces many of the aspects that characterized the apartheid workplace regime, such as job insecurity, low pay, and a lack of benefits.

From South Africa, the discussion of neoliberalism then moves to Aotearoa/New Zealand and Australia as Shaun Ryan and Andrew Herod explore how changes in the architecture of state regulation of labor markets have impacted the cleaning industry. Specifically, they outline that whereas for most of the 20th century the economies of both Australia and New Zealand were regulated through a system of industrial "awards" in which the state guaranteed certain minima of wages and work conditions as a way of ensuring both industrial peace and a modicum of social justice, more recently in both nations there has been a dismantling of this system so as to provide greater "flexibility" in the labor market. Though the timing of this dismantling differs in the two countries, its mark is indelible in both New Zealand and Australia and it is having significant consequences for cleaners. Specifically, with the abolition of government arbitration courts in New Zealand and the severe restriction of their ambit in Australia, labor markets have been "modernized"—which is to say they have been made more flexible. The result of this has been both a significant increase in the number of small and non-union cleaning firms but also the growing use of non-standard employment contracts facilitated through such means as outsourcing of work, with the result that cleaners have seen their wages and conditions of employment worsen significantly. Furthermore, as a result of government changes in the regulatory framework within which cleaners toil, it has become much more difficult for unions to organize to protect their interests, although, as Ryan and Herod recount, cleaners' unions in Australia and New Zealand have recently joined with the US Service Employees' International Union (sponsor of the Justice for Janitors campaign) to develop a tripartite organizing plan linking North America with Australasia.

In the final paper of this section, Patricia Tomic, Ricardo Trumper, and Rodrigo Hidalgo Dattwyler discuss how neoliberalism followed on the heels of the 1973 American-supported counter-revolution against Salvatore Allende's elected socialist government in Chile. Under the dictator Augusto Pinochet, neoliberalism moved from the classrooms of the Departments of Economics at the University of Chicago and the Catholic University of Chile—which was once described by José Piñera (nd), one of the architects of Chilean neoliberalism and an acolyte of Milton Friedman, as, intellectually, "a 'wholly owned subsidary' of the University of Chicago"—and into the offices of the planners of the Chilean economy. Here, neoliberalism was presented as the vehicle for the development of a "new, modern, and civilized" Chile to be realized through the privatization of economic activities, the promotion of deregulated and flexible labor markets, an increased reliance upon contingent work, and the growing penetration of foreign capital into the country. Significantly, as with the case of Thailand explored by Alyson Brody in Section 2, the neoliberals made a conscious effort to connect notions of modernization and development with those of hygiene and cleanliness. The result, Tomic et al point out, was the creation of a network of *cordons sanitaires* which connected the spaces of the modern economy and served to mark spatially the lines of division between "modern Chile" (represented by the shopping malls and educational institutions which serve as cocoons for middle class safety and security) and "*el otro*" ("the other") Chile—the "primitive" and "backward" spaces inhabited by those left behind by the Chilean neoliberal "miracle". Ironically, however, as Tomic et al show, it is the residents of this *otro Chile* who constitute the labor force who must keep the spaces of modernity sanitary and free from refuse if the image of neoliberal progress is to be maintained, even as they themselves must remain invisible lest their presence shatter neoliberal fictions.

References

Aguiar L L M (2001) Doing cleaning work "scientifically": The reorganization of work in the contract building cleaning industry. *Economic and Industry Democracy* 22(2):239–270

Brenner N and Theodore N (eds) (2002) *Spaces of Neoliberalism: Urban Restructuring in North America and Western Europe.* Malden, MA: Blackwell

Carroll W (2005) Introduction: Social democracy in neoliberal times. In W Carroll and R Ratner (eds) *Challenges and Perils* (pp 7–24). Halifax: Fernwood Books

Castree N (2005) The epistemology of particulars: Human geography, case studies and "context". *Geoforum* 36(5):541–544

Friedman T (2005) *The World is Flat: A Brief History of the Twenty-First Century.* New York: Farrar, Straus and Giroux

Keil R (2002) "Common-sense" neoliberalism: Progressive conservative urbanism in Toronto, Canada. *Antipode* 34(3):578–601

McBride S (2005) *Paradigm Shift.* 2nd ed. Halifax, NS: Fernwood Press

Ohmae K (1990) *The Borderless World: Power and Strategy in the Interlinked Economy.* New York: HarperBusiness

Piñera J (nd) *Chile: The Road to Freedom.* http://www.josepinera.com (last accessed 14 December 2005)

Chapter 2

Janitors and Sweatshop Citizenship in Canada

Luis L M Aguiar

Introduction

As neoliberalism has become ever more entrenched in Canada, the nation's janitors have increasingly come to face many of the same issues as do other workers, but they also face a number of quite specific concerns. Thus, on the one hand, like many workers they suffer lowered wages, the intensification of work, growing job insecurities, labor market precariousness, and having to take on increased responsibilities without corresponding increases in compensation. On the other hand, unlike many other workers, they face these issues at a distinct disadvantage since, for the most part, they remain largely "invisible" workers and generally lack union representation as a basis from which to mobilize and resist the worsening of their work conditions. These issues are stubbornly resilient and continue to torment janitors' lives, with more pain predicted for the future (MacDonald 1997), regardless of the political stripes of the government in power (Carroll and Ratner 2005; Panitch and Swartz 2003). In fact, as I shall argue here, neoliberalism, privatization, contracting out, and the lack of a comprehensive response to them (Gindin and Stanford 2003) have entrenched janitors in sweatshop conditions and eroded their ability to enjoy fully the citizenship rights which the institutions of the Fordist welfare state had previously extended to some and promised to others. Hence, beyond the workplace janitors suffer from the consequences of the dismantling of the welfare state, the privatization of public services, and the spread of workfare programs, with the result that they find themselves under attack in both the productive and reproductive spheres of their lives, an attack which limits their abilities to enjoy rights of citizenship which have long been assumed to be inviolable (Brodie 1996). This combination of persistent labor market precariousness, fast-eroding industrial rights, and a truncated social citizenship status for janitors brought about by the state's rolling back of many of the social and industrial rights that were extended to the working classes in the middle of the twentieth century is what is at the core of the concept of "sweatshop citizenship" which I explore here.

Neoliberalism in Canada

If the first epoch in Canada's effort at encouraging social and economic development began with the country's confederation in 1867 and was characterized by Prime Minister John A. MacDonald's "national policy" of connecting the country from coast to coast and of erecting tariff walls to protect its nascent industrial economy from competition by the United States, and if the second such epoch may be seen to have its origins in the interwar period and was marked by the Fordist compromise which saw the expansion of public sector work, the establishment of the welfare state, and the implementation of macro-economic policies based upon a Keynesianism which privileged theoretically and analytically the national economy (Radice 1984), then, arguably, the rise of neoliberalism at the end of the 20th century can be seen to represent the third great moment in the country's social and economic trajectory (Carroll 2005).[1] Significantly, however, the emergence of neoliberalism marks a noteworthy departure from what has gone before. Hence, the goal of federal government social and economic policy until recently has generally been to encourage a sense of national economic integration by ensuring that people enjoyed certain nationally protected workplace and social welfare rights by dint of their Canadian citizenship and/or residence. Of course, such "national" policies were always shaped by the particularities of class and regional struggles and compromises, and so often played out quite differently in different places. Nevertheless, there was an abiding sense that the role of the federal government was to establish a coast-to-coast set of standards (particularly with regard to the workplace) which would set an example for provincial governments and employers to emulate, although the latter were free to develop standards and practices which provided greater protections than those required by Ottawa. Such national standards, nonetheless, provided the foundation for fully enjoying the rights of citizenship.

By way of contrast, the driving ideological force of neoliberalism is the dismantling of national- and even provincial-level standards—which are seen as limiting the operation of market forces by preventing communities and workers located in different places from fully competing with each other by engaging in wage and regulatory whipsawing—and their replacement with an atomized economic and regulatory landscape which, it is argued, will allow market forces to function more efficiently. Certainly, as with Canada's earlier social and economic development projects, the implementation of neoliberalism itself is not a geographically monolithic process and its realization is occurring in a spatially uneven manner, the result of "the legacies of inherited institutional frameworks, policy regimes, regulatory practices, and political struggles" (Brenner and Theodore in Keil 2002:582) which continue to shape the politics of instigating new political projects in different places (Albo 2002; Mitchell 2004). However, a central element in the pursuit of neoliberalism and the contemporary "continental rationalization" of the country and its regions is the deliberate weakening of many of the institutions of "national" social and economic integration as a way of stimulating competition across the Canadian space-economy. Facilitated by Canada's federal system of "divided sovereignty", wherein provincial governments are responsible for many aspects of health, education, welfare, resource management, and labor law, this process is resulting in "an intensification of uneven development, as the relative fortunes of regions increasingly depend on how each is integrated into . . . continental and global markets rather than the national market" (Carroll and Little 2001:38–39).

The introduction of neoliberal policies, then, is dramatically transforming Canadian society in a number of ways (McBride 2001; Mitchell 2004). Thus, although the provinces and territories have always maintained significant control over much of what goes on within their jurisdictions, even as the federal government historically sought to equalize many standards across the country, neoliberalism is witnessing the devolution of much federal power to the provinces and territories, who have gained greater control on many social and economic issues and whose governments have increasingly seen their roles as ones of furthering provincial and territorial, rather than national, economic and social goals. Indeed, as it turns out, "much of the transition to neoliberalism has [in fact] been the handiwork of these subcentral governments" (Carroll and Little 2001:39). As part of this, provincial and local governments have increasingly turned away from playing the role of regulators of the economy and have adopted more entrepreneurial stances, offering various tax and other incentives attractive for business as a way to stimulate competition. These initiatives have been buttressed by an ideological attack on the welfare state, one which has been characterized by a "rhetoric of national deficit and decline, a decline that [has been] firmly linked [discursively] with the 'excess' of welfare state provision under [Prime Minister Pierre] Trudeau's [national] liberal government" (Mitchell 2004:131). This rhetoric has been used to create a sense of "moral panic" wherein Canada's Fordist economic practices and the social safety net of the welfare state were argued to be leading to the country's economic and social decay, such that only drastic measures could reverse its decline. The acceptance of such a rhetoric by substantial numbers of voters has made it possible for the forces of neoliberalism to replace consent with coercion in a number of social arenas during the past decade. Hence, whereas the fordist compromise of the mid-twentieth century relied upon Keynesian macro-economic policy and the social and workplace protections offered by labor unions, the "norms of free collective bargaining [have increasingly been] replaced with repressive measures discouraging the formation of unions, monitoring their practices, limiting their collective rights and (for public-sector workers) restricting or eliminating wage increases" (Carroll 2005:16). The success of this ideological assault is marked by the fact that by the early part of this new century neoliberal policies were being implemented across the country.

In the case of Ontario, neoliberalism has most clearly been manifested in the "Common Sense Revolution" of the rightist Conservative government of Mike Harris, which was elected in 1995 on a platform of welfare reform, health care restructuring, major reductions in public spending, and reduced income taxes (Keil 2002:588). Significantly, one of the provincial government's first pieces of new legislation was the "Labour Relations and Employment Statute Law Amendment Act (1995)", commonly known as Bill 7, which rewrote labor legislation pretty much for the specific purpose of showing capital that "Ontario was open for business" and that neither unions nor cumbersome labor law were going to stand in corporations' way. Thus, amongst other things, Bill 7 terminated bargaining rights for professionals who had previously been covered by it, it eliminated the province's card-based check-off certification system and replaced it with a vote-based scheme, it reduced the level of support required to secure a vote in a union termination application, it stipulated that a collective bargaining agreement does not take effect until it is ratified by a vote of the employees in the bargaining unit, and, except in the construction industry, it made any strike unlawful unless a strike vote is held and a majority of employees support it, all measures seen as limiting the power of unions. With regard to janitors, the Bill rescinded "successor rights", which had the immediate effect of putting their job security in doubt by making it easier for contractors to replace them, of bringing organizing to a virtual halt, and of sending wages into free fall (Aguiar 2000:82, table 5).[2] Indeed, such was the extent and depth of the Bill that some commentators have argued that it threw labor relations in the province back some 50 years (Schenk 1997).

On the other side of the country, in British Columbia, the centrist Liberal government's neoliberal platform came in the form of a 2001 "New Era" for business and the public sector, an era which promised to bring, amongst other things, reduced bureaucracy, privatization of many state-owned enterprises, greater opportunities for entrepreneurialism, welfare reform, making education more responsive to market demands, cuts in income taxes, and the introduction of "flexibility" into labor relations by limiting trade union rights and abolishing regulations which supposedly made the province's workers "uncompetitive". Hence, Bill 18 changed the process of union certification, such that, whereas previously a union which signed up 55% or more of eligible workers was automatically certified, now each certification application will have to undergo an election, a change which will drag out the process of emplacing a union in the workplace. More significantly, Bill 42 included a provision allowing any company that has been "inactive for two years" to decertify its union, thereby allowing it to declare bankruptcy and re-open for business under a different name and as a non-union operation (Nuttal-Smith 2002:A4). Moreover, the Bill also allows bankrupt companies to decertify before an ownership change, thereby making it more attractive for companies interested in being taken over or in selling off part of their operation to shift assets so as to declare bankruptcy and break their unions. Finally, it allows employers to communicate directly with workers in advance of a certification vote, a provision which previously would generally have resulted in automatic certification but which now gives employers a legal way to present workers with antiunion messages (Beatty 2002:A1). Not surprisingly, business groups are very pleased with these changes.

In effect, then, the Liberals have sought to ensure that the province's economy will be "reborn" without organized labor (Tieleman 2002) by "modernizing" the labor code to provide employers with "safeguards" and present employees with "incentives" to work harder. Such modernization has included increasing the power of the provincial labor board to declare workers "essential employees" (and thus ineligible to engage in certain union activities, such as strikes) and adding the promotion of corporate "competitiveness" as an aspect of the code's objectives, a provision widely seen as giving capital a means to justify further changes in work rules and job security provisions. According to the government, this introduction of flexibility and the encouragement of entrepreneurialism will be a "win–win" recipe for workers and their bosses as they increasingly find themselves having to compete in the global economy. However, early research on the effects of the government's labor code amendments shows that for those most vulnerable workers—such as janitors—the "New Era" of the 21st century has, in fact, undermined old safeguards and already begun to worsen workers' lot (Aguiar 2004; see also Cohen in this collection).

The Return of the Sweatshop

Neoliberal restructuring, then, is bringing with it deteriorating working conditions and labor rights for workers across Canada. Indeed, many writers have argued that such deterioration, combined with the expansion of informal work and of certain types of labor practices, means that growing numbers of Canadian workers are coming to experience what Munck (2002:4) and others view as the "brazilianization" of workplace conditions and relations, by which he means that the precarious forms of employment associated with informal work have now "become generalized" as "the result of the neoliberal free-market utopia unleashed by globalization". Such a "Third Worlding" of labor markets, however, has its origins not in the global South but in the practices of First World capitalism, practices which were largely hidden from view during the Fordist compromise and welfare statism of the twentieth century.[3] Thus, in his analysis of contemporary trends in Los Angeles—a city which he sees as "a paradigm of First and internal Third Worlds"—Sawhney (2002:2) argues that the spread of neoliberal practices and the subsequent contemporary restructuring of the capitalist economy have largely "exteriorized" and generalized the previously hidden exploitative and abusive conditions suffered by a marginal "invisible workforce" onto the labor experiences of an increasing number of workers who formerly enjoyed the benefits and securities of Fordist work arrangements—that is to say, those who were unionized, those who were engaged in long-term work relationships with their employers, and those who were full-time workers. Of particular interest in this regard is the fact that one key aspect of the "new economy" which is emerging under neoliberalism has been the spread of sweatshop conditions for many workers, an aspect of capitalist production that was often assumed to have been long eradicated. This spread of sweatshop conditions is leading to the growth of what I will call here "sweatshop citizenship".

The "rediscovery" of sweatshops within Canada—often through high-profile media exposés—has shocked many people, for part of the success of the Fordist state was its apparent elimination of sweatshop labor conditions via an expansion of "industrial citizenship" and the uplifting provisions of the welfare state (Basok 2002; Ng 1998). Thus, under Fordism union contracts were typically structured so that higher productivity would lead to higher wages and, presumably, a better standard of living, all of which was overseen and managed by a corporatist state. In addition, the pursuit of safe working conditions, good terms of employment, industrial rights and recognition, and the acceptance of unions were viewed as legitimate and things the state should protect. Indeed, as fordism developed and solidified itself, the public's collective memory generally increasingly located sweatshops in a rapacious 19th century capitalism which had produced the satanic mills about which both William Blake and Karl Marx had written. By the late 20th century, then, although the Canadian public was generally aware of the continued existence of sweatshops in developing

countries—the result, amongst other things, of well-publicized campaigns against the use of children in Indian carpet factories and the
public shaming of television personality Kathy Lee Gifford (whose
clothing company was found to be exploiting teenage girls in
Honduras)—most believed that domestically sweatshops were historical artifacts (Maquila Solidarity Network nd; Ng 1998).

It is not just the resurgence of sweatshops and the fact that "they
are right here" (Ross A 2004:10) which has come as something of a
shock to many but, rather, their ubiquity across the post-Fordist/
post-industrial economic landscape. Hence, sweatshops are not just
found in those urban ghettos in which migrants from the Third world
live in Canada, nor are they only to be found in those industries in
which they have typically flourished, such as garment manufacturing.
Rather, they are increasingly common in industries as diverse as
footwear and toy manufacturing, construction, food preparation and
car washing, the paid domestic work sector, the home renovations
industry, in auto repairs, in landscaping, and in janitorial services,
as well as in many other parts of the supposedly "high-tech, post-
industrial" economy. Furthermore, sweatshops are not only found in
grim innercity industrial ghettos, but also in the countryside in
meatpacking plants and farms, and have also begun to appear in the
suburbs (Gordon 2005:2).

Certainly, not everyone accepts that there has been an explosion in
the number of sweatshops across the Canadian landscape. Robert Ross
(2004:26), for example, has argued that much of the perceived
"expansion" of sweatshops is the result simply of too broad a
definition of what constitutes such a workplace. Consequently, he is
skeptical of those who want to categorize as sweatshop work work which
is simply "lousy", preferring, instead, a narrower and more legalistic
definition which sees a sweatshop as "a business that regularly violates
both wage or child labor and safety or health laws". For her part,
whilst she shares similarities with Robert Ross and generally tends to
see as sweatshops those places of work with illegally long hours, low
wages, and high rates of injury, Gordon (2005:13–14) argues that
we should not think that there is some natural economic barrier which
"keeps sweatshop conditions . . . in traditional sweatshop industries"
such as garments and so should not be surprised when we see them
emerging in other industries. Rather, she suggests, as jobs in general
have shifted from manufacturing to the service sector, so "sweatshops
have [simply] followed suit", with the result that although today's
"[s]weatshops may look different [from] their predecessors of an
earlier era", their effect on workers is "often devastatingly the same"
(Gordon 2005:14).

By way of contrast, Andrew Ross (1997, 2004) suggests that much
of the debate over whether a broader or a narrower definition of

sweatshops is most appropriate largely misses the point because, despite their differences, both those who argue for a narrower definition and those who argue for a more expansive definition still tend to conceive of sweatshops chiefly in terms of whether or not laws are actually being broken—that is to say, they tend to view workplaces as more eligible for the designation of "sweatshop" the less they comply with prevailing law. In adopting such an approach, however, these commentators fail to see that the nature of the law has itself changed, that, in other words, the point of reference has been moved by neoliberal-inspired legislative actions. Thus, he points out, restricting the definition of a "sweatshop" to that of a business which violates (whether routinely or not) labor law or health and safety laws fails to recognize that much of the exploitation and abuse which was previously illegal is now actually legitimized by the new legislation being passed as governments neoliberalize the workplace. Hence, he maintains (2004:22, emphasis added), "most low-wage jobs, *even those that meet minimum wage requirements and safety criteria*, fail to provide an adequate standard of living for their wage earners, let alone their families". For him, then, a better definition of whether establishments are sweatshops revolves not around whether their practices are legal or not but, rather, whether they are firms that provide a "living wage" (see also Bonacich and Appelbaum 2000). For Ross:

> In most respects, it is the systematic depression of wages, rather than conscious attempts to evade labor laws, that is the structural problem. Installing proper fire exits may turn a sweatshop into a legal workplace, but it remains a low-wage atrocity. All the more reason to define and perceive the "sweatshop" as a general description of all exploitative labor conditions, rather than as a subpar outfit, as defined by existing laws in whatever country the owner chooses to operate. (Ross A 2004:22–23; see also 1997:296)

There are two advantages, I want to argue here, to adopting this latter definition of sweatshop. First, it moves debate about the emergence of sweatshops beyond a crude dualism of whether or not firms are simply breaking the law. This is important, since much of the exploitation and poor working conditions which workers face is actually quite legal under contemporary labor legislation and is becoming more so as neoliberalism penetrates ever greater parts of the economy. Put another way, adopting a definition which centers upon the absence of a "living wage" and good working conditions prevents governments from "defining away" the problems of sweatshops by simply legalizing what were previously illegal activities. Second, such a definition shatters any delusions that sweatshops are things found only in the "Third World", far away from the modern capitalist economy. Thus, if we are to take neoliberals at their word—namely that the removal of government regulations will allow the market to operate "as it should", without the distortions brought about by political interventions—then the contemporary growth of low-waged work and poor working conditions which market deregulation is auguring in Canada and elsewhere is not actually an aberration but is, rather, emblematic of the nature of unfettered, "free market" capitalism. This puts paid to the notion that the pursuit of neoliberal deregulation will bring about the vaunted "high wage, high skill" economy about which free market advocates so often talk (see, for example, Reich 1991).

In the case of janitors, it is already evident that neoliberal labor market deregulation is threatening to turn the workplace very much into a 21st century sweatshop (Canadian Union of Public Employees 1997; Ehrenreich 2003; Giles 1993; MacDonald 1997; Production Multi-Monde 1991). This is significant, because although the cleaning industry is generally not thought of as being an exemplar of the "high wage, high skill" economic model which neoliberals argue will result from their policies, as Sassen (1990) has argued, neither is it simply a remnant of some bygone era of capitalism. Indeed, professional cleaning is one of the fastest growing employment sectors in both North America and Western Europe, for janitorial services are, in fact, central to the functioning of the contemporary global(izing) economy, whether it is in making office space serviceable or in cleaning the homes and playscapes of the new bourgeoisie in the global cities which serve as control points of the planetary economy. And yet, for all its importance for the operation of contemporary capitalism, much of the industry is developing in spaces where sweatshop conditions are being institutionalized in the daily operations of the companies who employ the millions of janitors who keep the manufacturing and office spaces of the global economy clean. In the process, janitors are being denied many of the rights which those in the industry previously enjoyed, a development which is subjecting them to practices of what I am here calling "sweatshop citizenship".

From Industrial Citizenship to Sweatshop Citizenship

It is, I think, fair to say that much of the debate today about citizenship engages at some level with the arguments laid out by sociologist Thomas Humphrey Marshall in works such as his 1950 book *Citizenship and Social Class*. For Marshall, full citizenship was constituted by three sets of rights: civil rights (which included freedom of thought and expression, the right to own property, and a right to "justice"); political rights (which included the right to participate in the political process through such things as elections); and social rights (which included the right to earn a living and to enjoy a level of social and economic security compatible with the standards prevailing in society). Most importantly, for Marshall these rights of citizenship could only achieve their fullest expression within a liberal-democratic welfare state operating within a capitalist market economy—that is to say, the potential extremes of capitalism had to be tamed by the regulatory framework and support of a welfare state if the citizenry were to be truly able to participate in the decisions which affect their lives.

Drawing upon Marshall's work, then, it is possible to contrast the concept of an "effective citizenship" with that of what we might call a "nominal citizenship" (Harrison 1995:20–21), with the former incorporating notions of citizenship as both "status" (which allows "enjoyment of civil, political and social rights" within one's country of citizenship) and as "practice" (which "requires the acceptance and performance of wider communal responsibilities and duties") whereas the latter merely refers to one's status as a member of a particular collectivity defined by the spatial boundaries of a given nation-state (being Portuguese or Chilean, for example) (Dwyer 2004:4). This distinction between effective citizenship and nominal citizenship is important, because under Fordism the widening and expanding of the welfare state was generally seen as a way of both facilitating the greater enjoyment of citizenship rights by individuals—it would be hard, for instance, to argue that individuals can fully exercise their rights as citizens if they cannot read and write or if they suffer such ill-health or poverty that they cannot participate in the public arena—and of ensuring that social actors such as firms lived up to the wider societal expectations of being, for instance, "good corporate citizens" by respecting particular communally-agreed-upon rights (but see, Park 2004; Siltanen 2002; Vosko 2000). In the case of the workplace, such ideas became manifested in the classic liberal pluralist concept of "industrial citizenship", wherein workers could be seen as engaged in carving out rights and obligations which, once established, would subsequently be attached to state functions. In such a concept, employers and workers were idealized as actors engaged in a "politics of production" (Burawoy 1985) which involved collectively negotiating the parameters of such industrial citizenship under the auspices of a protective state which would both provide the framework within which the process took place and subsequently enforce legally what was agreed upon.

Today, in contrast, such ideas of citizenship are being undermined. Hence, whereas in the past the state expanded the social rights of citizenship, it has recently attacked such rights, clawing them back and, in the process, restructuring what Canadian citizens have come to expect as their "rights of citizenship" (such as free and universal healthcare and education) (Radcliffe 2005:324). Thus, many of the gains of the Fordist period (such as restrictions on the length of the workday and the idea of a minimum wage) are under attack, leading some to talk about the emergence of "post-industrial citizenship" rights (Arthurs 1996). In such a world, individual rights are increasingly privileged over collective rights, such that a "variable geometry of citizenship" (Calderón, Assies and Salman 2002:15) appears to be replacing the calculus of universal citizenship rights. This transformation of the Fordist regulatory environment is being accomplished in two manners. First, despite the neoliberal rhetoric about the need for smaller government and market deregulation, the undermining of industrial citizenship rights is being brought about not so much by a withdrawal of the state from the public arena (as the term "deregulation" would suggest) as it is, frequently, actually being brought about by a new form of intervention—a process marked by re-regulation rather than deregulation, whereby, for instance, greater requirements are placed on the unemployed to be able to claim benefits or regulations are introduced making it harder for unions to call strikes (Keil 2002).

Second, the regulatory burden is being passed from the public to the private sector—rather than regulating directly, the state is transforming the regulatory environment so as to regulate via surrogates. Thus, for instance, the re-regulation of the British Columbia (BC) forestry industry by the Liberal government has taken the form of encouraging "peer pressure", whereby employers are left to pressure one another to comply with their own concept of proper business practices, rather than ensuring compliance with regulations via government inspections. Indeed, as part of this new model of regulation the government has actually laid off numerous workplace inspectors, with the, perhaps, not surprising result that deaths in BC forestry work rose to 41 in 2005, a number which is 16 (or 64%!) greater than the annual average over the last 80 years (Kennedy 2005). A similar attitude can be seen in the Bush administration's approach to regulating pollution, wherein it has argued that instead of government regulators doing the job of keeping air clean, dismantling regulations will allow the market to work more efficiently, with consumers punishing polluters and rewarding cleaner companies through their purchases.

The strength of Arthurs's concept of post-industrial citizenship, then, lies in its ability to capture the wholesale changes taking place in the Canadian industrial relations regime, changes which have been quite fundamental and which have significantly increased workers' anxieties (Panitch and Swartz 2003). However, its strength is also its weakness, for in being an all-encompassing concept it cannot effectively grasp that workers are not impacted equally or with the same degree of severity. For instance, the idea of post-industrial citizenship does not capture the division between primary and secondary labor markets, nor ethnic and gender divisions, even though we have ample evidence showing that those in the secondary labor market are treated more poorly by their employers, earn much less, and are often still ignored by the labor movement (Ehrenreich and Hochschild 2003; Gabriel 1999; Giles 2002; MacDonald 1997; Waldinger and Lichter 2003). Nevertheless, it does, I think, serve as a useful marker for signaling a significant change in workplace relations. Thus, although under fordism some workers—women, immigrants, visible minorities —often suffered from an "incomplete citizenship" due to their location in the labor market and their frequent ignoring by the labor movement and state actors, the inclusive ideology of Fordist industrial citizenship nonetheless envisioned that, at some future point in time, they would enjoy the fruits of full membership. On the other hand, with post-industrial citizenship there is no longer that same commitment to an equity agenda in the state's programs and policies, with the result that it is implicitly understood that effective citizenship will be indefinitely deferred for workers such as janitors. Put another way, as the economic role of the state has been redirected from one of ensuring minimum labor standards to one of encouraging entrepreneurialism within the context of the shift to a regime of global neoliberalism, the state's previous role as an agent which seeks to ensure that individuals can exercise fully their citizenship rights is "no longer [as] central as communities are [being] reorganized based on [their] relations to the global system" (Tabb 2001:56).

The transformation in the state's political economic role has been matched by a neoliberal rhetoric which suggests that labor market inequality is individually based rather than structurally anchored, a rhetoric in which economic "winners" are exalted but workers such as janitors are invariably condemned as economic "losers", stuck at the bottom of the labor market as a result of their lack of education and/or willingness to work hard. Somewhat insidiously, this rhetoric has been internalized by many unionists, who have moved away from emphasizing long-standing models of inclusive industrial citizenship and have increasingly felt pressured to prepare workers individually for a competitive labor market through such things as job retraining schemes (Gindin and Stanford 2003). Hence, whereas "the discourse of [industrial] citizenship once provided the basis for an inclusive vision of unions' constituencies, the language of competitiveness and productivity is [now] undermining labor's ability to develop collective demands, collective identity, or collective action" (Seidman 2004: 118–119). This presents tremendous challenges to those who would seek to defend the lot of workers such as janitors who have seen themselves increasingly emplaced on the margins of the labor market, their wages systematically depressed and their jobs made permanently insecure as they are forced to endure the erosion of the civic, political, and social rights which Marshall suggested were characteristic of "effective citizenship". Indeed, so bad is the position of such workers becoming that if the concept of a post-industrial citizenship is useful for distinguishing generally between the practices of the Fordist and of the post-Fordist era, such that even today's economic "winners" are seeing many of their workplace rights eroded, then it is, perhaps, not inappropriate to describe the economic "losers" who toil in the burgeoning sweatshops of the Canadian economy as enjoying not so much post-industrial citizenship rights but, rather, forms of "sweatshop citizenship". In what follows I will outline how such sweatshop citizenship is being entrenched in some labor markets and workplaces.

Entrenching Sweatshop Citizenship

> It's worse, way worse. If you figure out the medical and dental, we get less. It's hard to find a job. What are you going to do? Where do I go? On welfare? It's pretty tight. [But] I'll [have to] take it. (Delfina Girardi, janitor at Simon Fraser University; Aguiar 2004).

Delfina Girardi made this remark after applying for the same cleaning job she had held for 17 years at Simon Fraser University when, in 1994, the University's Board of Governors awarded the campus cleaning contract to Marriott Corporation of Canada rather than retain its long-standing cleaning service, Empire Maintenance Industries, Inc. The choice was easy for the Board, since Marriott offered to clean the university for C$296,480 less than what Empire had proposed. This "saving" was gained entirely by Marriott's under-cutting of janitors' wages, from Empire's rate of between C$9.49 and C$11.50 an hour to between C$7.50 and C$11.01. More importantly, Empire's failure to retain the contract meant that janitors were immediately terminated and left without a job. Consequently, janitors had two choices: find work elsewhere or reapply for the same work they had done for years, but at two dollars less per hour! Delfina, like many of her workmates, reapplied for her job because it was the only thing she could do to secure an income. For its part, Marriott was able to engage in such practices because of a clause in British Columbia's labor legislation which ensures that contractors have no obligation to existing workforces when taking on a contract, a provision which is symptomatic of the difficulties that contracting out imposes on building janitors in British Columbia and elsewhere.

Such contracting out of work—in combination with the lack of a successorship clause for workers—is frequently used systematically to depress wages. Thus, in a recent analysis of the privatization of hospital janitorial services in Victoria, BC, Nelson (2005) found that administrators deliberately used this as a strategy to keep costs down, with the result that it is the janitors themselves who bear directly the costs of outsourcing—in the case of one worker, for instance, the outsourcing of cleaning resulted in a C$5 per hour wage cut, such that, after 23 years at the hospital, the worker was earning only C$11.79 per hour, a rate far below a living wage in Victoria, which has some of the highest housing costs in the country. Similar patterns can be found in other regions such as Toronto, where immigrant women janitors were earning on average C$17,000 a year in the 1990s, so low that many ended up working at two or more jobs just to make ends meet (Giles 2002:8).

Such employer efforts to reduce janitors' wages have been exacerbated through the practice of "double-breasting", a practice in which one company will have multiple sub-divisions which it manipulates to keep the price of labor down. As one worker explained:

> They have two companies: the E.T.D. Building Tenants and the Priority Building Services. These are owned by the same owner, same management. But the E.T.D. concentrates on those contracts that they get and provides the labor supply, whereas the Priority (which they own also) is the company that takes charge of the franchising. So when they get contracts from the public, if they cannot sell this to the public through franchising, then they handle it themselves through the E.T.D. Building Tenants. If they can sell it through franchising, then they use the Priority Building Services . . . [This way Priority doesn't] pay for any labor, they just provide the training and supplies and in return they get royalties, something like 20% monthly. (Author interview with Wilfredo Bagunu, 26 June 2001)

In such a business strategy, then, companies will place their subdivisions into direct competition with each other as a way to underbid competitors' or even their own bids once contracts have been tendered, with the result that janitors often end up competing against themselves. Contracting-out and double-breasting, then, are significantly reducing janitors' income and making it difficult for them to enjoy many of the rights of citizenship which they had previously come to expect—low incomes and having to work two or more jobs have implications for their and their children's health and education, their ability to participate in the broader activities of civic life, and so forth.

In the janitorial labor process, sweatshop conditions are evident in the work intensification practices janitors endure as management restructures work. For them, the restructuring of work has meant harder, faster, and more strenuous work but also consistently declining work time and stagnant or reduced wage rates. Hence, in Toronto, janitors typically used to clean one office floor in a seven-hour work shift. Today, they clean one-and-a-half or two floors in a six- or even five-hour shift, with no noticeable decline in tenancy rates in the buildings which they clean. Likewise, in Place Ville Marie office and shopping complex in Montreal, a Portuguese cleaner reported that janitorial management had reduced the workforce from 350 to 150 janitors and yet demanded from each who remained an increase in their cleaning zone of 800 ft^2 (Aguiar 2001:252)! Significantly, these changes have come hand-in-hand with contracting out and the adoption of the Total Quality Management (TQM) mantra in which janitors are subjected to a computerized program (known as the Specialized Maintenance Management System) which conducts time and motion studies of cleaning tasks to measure the "proper" amount of time allotted to any particular task, with this information then being used by management to assign the "correct" number of workers required to clean a particular area. Not only does this system intensify the labor process, but it also reduces janitors' autonomy in the work-place, autonomy which had previously allowed them to have some measure of control over the pace of work and to use their own ability and skill to determine what constituted a "clean" office (Aguiar 2001).

If contracting-out and the adoption of TQM techniques have allowed employers to reduce their wage obligations and intensify work, the lack of successorship clauses, especially in the service sector (Cohen 2002), has greatly undermined workers' job security because workers potentially face dismissal every time a contract comes to an end, as any new contractor is free to bring into the workplace their own crew. Indeed, as far as I am aware, there is, in fact, no provincial labor code which contains legislation stipulating that a new contractor must retain the existing workforce after their original contractor has lost a work bid (Aguiar 2000; White 1993) (see Ryan and Herod in this collection for an account of how some Australian unions have tried to enforce "status quo" provisions to provide some job security). Given the greater spread of the practice of contracting-out in the professional cleaning sector relative to many others, the lack of such clauses means that janitors are exposed to the vicissitudes of neoliberalized labor markets much more than are many other workers, for they constantly face the threat of job loss as corporations and public employers relent-lessly seek ever cheaper contracting-out arrangements.

Moreover, such labor market practices perpetuate janitors' vulner-ability, for the lack of successorship rights not only institutionalizes job insecurity but also impedes janitors' abilities to organize cam-paigns for union representation in two ways. First, unless janitorial unions can organize all cleaning firms simultaneously and keep them organized, then those firms where they are successful in raising wages and securing benefits are likely to be uncompetitive when bidding for work relative to contractors who use unorganized and less expensive workforces, such that unionized janitors are liable to lose their jobs when their current employer loses a contract. In such a deregulated labor market in which work is increasingly awarded on the basis of competitive subcontracting, the existence of even a relatively small handful of non-union companies can quickly serve as a wedge to encourage others to go non-union or face always losing bids on contracts—a de-unionization made easier by the lack of successorship clauses. Second, the relentless turnover of janitorial workforces as contracts are won and lost makes it difficult for janitors to establish many of the kinds of workplace links of friendship and camaraderie which take time to mature but which are necessary for the development of the practices of solidarity out of which effective unionism can grow.

Certainly, I do not wish to suggest that there have not been rays of hope for janitors. Thus, the election of the New Democratic Party (NDP) in Ontario in 1991 did bring about some changes to the province's labor code late in the NDP's first and only term in office, including a clause whereby successorship was extended to service workers and janitors. The effects of this were immediate and highlight the importance of successorship clauses for unionization. Hence, not only were janitors protected—they kept their jobs—when another contractor took over the building they were employed to clean, but union organizing dramatically increased, and even a pay equity plan was rumored to be afoot (Aguiar 2000).[4] However, in 1995 the NDP was defeated and the new Conservative government proceeded almost immediately to re-write the provincial labor code, rescinding the successorship provision for janitors. The impact of this was immediate—organizing of new workers ground to a halt and protecting gains earned became unions' priority (Aguiar 2000; Panitch and Swartz 2003:235).

Whilst the NDP's tenure in Ontario was relatively brief, in British Columbia the party ruled for over a decade. Nevertheless, whereas the government did make several changes to the labor code, successorship rights were never included in provincial legislation (MacDonald 1997). Although the NDP did try to include such rights and extend their reach through a number of "sectoral certification" proposals (wherein bargaining units would be merged and standard contracts would be negotiated to apply to all work done in particular sectors), it was savagely attacked by business interests, to the extent that its own government members lost their nerve and succumbed to opposition pressure. Thus, as Glen Clark (2001), the former leader of the NDP and Premier of BC, has explained:

We wanted to bring in successorship rights so that, basically, if a contractor unionized they would stay unionized, even if the contract was lost to another company. But . . . it is very hard to effect controversial change when you are unpopular. It's hard psychologically and it's hard in reality. We had a one-seat majority in the legislature, we were at 20% popular support, and we faced very hostile media. We are talking about getting elected again. Most politicians only care about getting elected again. They look at the polls, so when you are at 20%, then the majority of politicians says "Holy shit, we are at 20%! We better not do anything to piss anybody off. We should do good things; try to get more support." Right? That never really affected my thinking particularly but that's certainly what everybody else thinks and, believe me, if you try and do something controversial and unpopular, the MLAs [Members of the Legislative Assembly] go crazy. They want to get re-elected, especially if they got into politics because they are into some interest group and they don't really have a class analysis or a philosophical or ideological compass, then when the shit hits the fan, they are not there. (Author interview with Glen Clark, 21 August 2001)

Even as opposition from business interests was intense, it is also the case that, given the difficulties that new legislation poses union organizers, some unions appear to have seen some advantages to maintaining a lack of successorship clauses. Thus, as Irma Mohammed (2003), a representative from the BC Federation of Labour, has argued, some union leaders appear to be practising a type of "enlightened self-interest" in their rather lackluster support for successor rights, seeing contracting out as "a way to pick-up organized members previously organized as members of other unions" (author interview, 13 May 2003). Evidently, just as individual capitalists may put their own interests above their class interests, so, too, do some unionists!

Despite such problems, though, some efforts are being made to organize janitors, although this is proving difficult, both because of the structural context within which such organizing must take place and because of the pressure on unions to get the best "return" on their investment of time and money spent organizing. Thus, whereas the labor movement recognizes that it must redouble its efforts (Glenday 2000), the fact that janitorial workplaces are typically small and janitors are amongst the poorest paid service sector workers means that even when campaigns are successful they generally deliver few new union members and those members' meager wages result in low union dues, which limits the resources available for organizing more workers. The result is that janitors remain marginalized in a left political agenda that can neither secure sectoral organizing measures in the labor code nor resist the neoliberals' policies which make it simultaneously more difficult for workers to unionize and easier for employers to decertify those unions which currently represent janitors. And, all the while, the neoliberal agenda becomes ever more entrenched in provincial labor codes.

Conclusion

What, then, does the future hold for janitors? Can they permanently eradicate their sweatshop citizenship status in Canada? If so, how? These are difficult questions to answer, for any course of action will require involving a significant mobilization of, and on behalf of, janitors, at a time when attacks on janitors' working conditions and levels of remuneration mean that many must work two or even three jobs just to make ends meet and generally have little spare time to devote to political or other activities. Yet, there are examples from elsewhere that may prove instructive. Perhaps the success of the Justice for Janitors campaigns in the US can offer a way toward resolving janitors' pressing issues in Canada (Erickson et al 2004), although so far there is little evidence of the JfJ presence in janitors' workplaces here, despite the fact that the Service Employees' International Union (SEIU) has many locals in Canada, some of which have at one time or another represented building janitors. At the same time, it is clear that the make-up of the Canadian labor movement's internal power structure is likely to require a significant transformation if it is to be successful in those sectors of the economy in which immigrants are predominantly employed, such as cleaning. Thus, whereas janitorial work in Vancouver and Toronto is almost exclusively done by workers from minority backgrounds, ethnic and visible minorities have, to date, enjoyed virtually no positions of influence within the movement's leadership (Ogmundson and Doyle 2002). However, such a necessary transformation is likely to cause significant disruptions within the labor movement as old lines of power are challenged and as new leaders with different agendas emerge (see Savage, this collection, for an example of how this played out in one janitors' union in Los Angeles).

What is clear, then, is that the Canadian labor movement's "confidence remains shaken by the hostile political-economic environment of neoliberalism" and that the movement has so far failed to "articulate a thorough critique of neoliberalism" (Gindin and Stanford 2003:438). Any such critique must include an examination of the forces bringing about "sweatshop citizenship". The challenge is clearly enormous. However, workers and organizers cannot ignore the situation within which they find themselves and must find a way forward that is inclusive of workers and their communities and that will allow them to secure processes to prevent reactionary governments from assailing progressive legislation once they are in power. The great unanswered question, however, is this: if janitors' wage and working conditions were bad under Fordism when there was at least a consensus across the political spectrum regarding an expanding industrial and social citizenship, how much worse must they get in the face of insidious neoliberalism, sweatshop citizenship, and the unrepresentative composition of the leadership in the Canadian labor movement before enough people say "enough is enough"?

Acknowledgments

I would like to thank the following people for their contributions in making this a better paper: Dave Broad, two anonymous reviewers, and Andrew Herod. Andrew has been especially insightful in his editorial capacity in giving this paper improved shape, scope and better focus in illuminating the central argument made herein. I would also like to thank Wilfredo Bagunu, Irma Mohammed and Glen Clark, especially, for giving me their time to discuss janitorial issues in BC. Research for this paper was supported by a grant from the Social Sciences and Humanities Research Council of Canada (SSHRC).

Endnotes

[1] The Fordist compromise refers to a system in which employers and unions agreed, essentially, to link wage rates and improvements in working conditions with improvements in productivity. For its part, the nation-state ensured certain basic minimum conditions of work for workers as part of its pursuit of demand-side macro-economic policies.

[2] Successor rights ensure that workers can continue to enjoy conditions of employment that were established with one employer even if their work is subsequently contracted out to another.

[3] Thanks to Dave Broad for reminding me that other writers have also made this point about the "Third Worlding" of the First World.

[4] It is important to realize, however, that even with the change introduced by this legislation, janitorial work still remained low paid and janitors continued to be some of the most vulnerable of workers (cf White 1993).

References

Aguiar L L M (2000) Restructuring and employment insecurities: The case of building cleaners. *Canadian Journal of Urban Research* 9(1):64–93

Aguiar L L M (2001) Doing cleaning work "scientifically": The reorganization of work in the contract building cleaning industry. *Economic and Industrial Democracy* 22(2):239–269

Aguiar L L M (2004) Resisting neoliberalism in Vancouver: An uphill struggle for cleaners. *Social Justice* 31(3):105–129

Albo G (2002) Neoliberalism, the state, and the left: A Canadian perspective. *Monthly Review* 5(1):46–55

Arthurs H (1996) *The New Economy and the Decline of Industrial Citizenship.* Kingston, Ontario: Industrial Relations Centre, Queen's University Press

Basok T (2002) *Tortillas and Tomatoes: Transmigrant Mexican Harvesters in Canada.* Montreal: McGill-Queen's University Press

Beatty J (2002) Labor law overhauled to give employers more flexibility. *Vancouver Sun* 14 May

Bonacich E and Appelbaum R (2000) *Behind the Label: Inequality in the Los Angeles Apparel Industry.* Berkeley: University of California Press

Brodie J (1996) Restructuring and the new citizenship. In I Bakker (ed) *Rethinking Restructuring* (pp 126–140). Toronto: University of Toronto Press

Burawoy M (1985) *The Politics of Production.* New York: Verso

Calderón M, Assies W and Salman T (eds) (2002) *Ciudadanía, Cultura Política y Reforma del Estado en América Latina [Citizenship, Political Culture and State Reform in Latin America].* Zamora, Mexico: el Colegio de Michoacán

Canadian Union of Public Employees (1997) *Contracting Out Custodial Services Destabilizes Edmonton's Public Schools. Contractors Receive Failing Grade.* Edmonton: Canadian Union of Public Employees

Carroll W (2005) Introduction: Social democracy in neoliberal times. In W Carroll and R Ratner (eds) *Challenges and Perils* (pp 7–24). Halifax: Fernwood Books

Carroll W and Little W (2001) Neoliberal transformation and antiglobalization politics in Canada. *International Journal of Political Economy* 31(3):33–66

Carroll W and Ratner R (2005) The NDP regime in British Columbia, 1991–2001: A post-mortem. Paper presented at Pacific Sociological Annual Meeting, Portland, Oregon

Cohen M (2002) Globalism's challenge for feminist political economy and the law. Paper presented at the Seventy-Third Annual Meeting of the Pacific Sociological Association. Vancouver, British Columbia

Dwyer P (2004) *Understanding Social Citizenship.* Bristol: The Policy Press

Ehrenreich B (2003) Maid to order. In B Ehrenreich and A R Hochschild (eds) *Global Woman* (pp 85–103). New York: Metropolitan Books

Ehrenreich B and Hochschild A R (2003) *Global Woman.* New York: Metropolitan Books

Erickson C, Fisk C, Milkman R, Mitchell D and Wong K (2004) Justice for janitors in Los Angeles and beyond. In P V Wunnava (ed) *The Changing Role of Unions* (pp 22–58). Armonk, NY: ME Sharpe

Gabriel C (1999) Restructuring at the margins: Women of color and the changing economy. In E Dua and A Robertson (eds) *Scratching the Surface* (pp 127–164). Toronto: The Women's Press

Giles W (1993) Clean jobs, dirty jobs: Ethnicity, social reproduction and gendered identity. *Culture* 13(2):37–44

Giles W (2002) *Portuguese Women in Toronto.* Toronto: University of Toronto Press

Gindin S and Stanford J (2003) Canadian labor and the political economy of transformation. In W Clement and L Vosko (eds) *Changing Canada* (pp 422–442). Montreal: McGill/Queen's University Press

Glenday D (2000) Off the ropes? New challenges and strengths facing trade unions in Canada. In D Glenday and A Duffy (eds) *Canadian Society* (pp 3–37). Toronto: Oxford University Press

Gordon J (2005) *Suburban Sweatshops.* Cambridge, MA: The Belknap Press of Harvard University Press

Harrison M L (1995) *Housing, "Race", Social Policy and Empowerment.* Aldershot: Avebury

Keil R (2002) "Common-sense" neoliberalism: Progressive conservative urbanism in Toronto, Canada. *Antipode* 34(3):578–601

Kennedy P (2005) Labour minister seeks to protect forest workers. *Globe and Mail* 6 December:S3

MacDonald D (1997) Sectoral certification: A case study of British Columbia. *Canadian Labour and Employment Law Journal* 5/6:243–286

Maquila Solidarity Network (nd) http://www.maquilasolidarity.org (last accessed 30 October 2005)

Marshall T H (1950) *Citizenship and Social Class*. Cambridge: CambridgeUniversity Press

McBride S (2001) *Paradigm Shift*. Halifax: Fernwood Books

Mitchell K (2004) *Crossing the Neoliberal Line*. Philadelphia: Temple University Press

Munck R (2002) *Globalisation and Labour: The New "Great Transformation"*. London: Zed Books

Nelson M (2005) "Janitorial or Light Duty? The Politics of Institutional Cleaning." Unpublished paper (copy in author's possession)

Ng R (1998) Work restructuring and recolonizing Third World women: An example from the garment industry in Toronto. *Canadian Woman's Studies* 18(1):21–25

Nuttal-Smith C (2002) Liberals "going too far", poll finds. *Vancouver Sun* 18 February

Ogmundson R and Doyle M (2002) The rise and decline of Canadian labor, 1960 to 2000: Elites, power, ethnicity and gender. *Canadian Journal of Sociology* 27(3):413–454

Panitch L and Swartz D (2003) *From Consent to Coercion*. Toronto: Garamond

Park J S W (2004) *Elusive Citizenship*. New York: New York University Press

Production Multi-Monde (1991) *Brown Women, Blond Babies*. Montreal: Video recording

Radcliffe S (2005) Neoliberalism as we know it, but not in conditions of its own choosing: A commentary. *Environment and Planning A* 37(2):323–329

Radice H (1984) The national economy—a Keynesian myth? *Capital and Class* 22:111–140

Reich R B (1991) *The Work of Nations: Preparing Ourselves for 21st Century Capitalism*. New York: A A Knopf

Ross A (1997) After the year of the sweatshop: Postscript. In A Ross (ed) *No Sweat* (pp 291–296). New York: Verso

Ross A (2004) *Low Pay, High Profile*. New York: The New Press

Ross R (2004) *Slaves to Fashion*. Ann Arbor: University of Michigan Press

Sassen S (1990) *The Global City*. Princeton: Princeton University Press

Sawhney D N (2002) Journey beyond the stars: Los Angeles and Third Worlds. In D N Sawhney (ed) *Unmasking LA* (pp 1–20). New York: Palgrave

Schenk C (1997) New organizing strategies. *Canadian Dimension* 31(4):15–16

Seidman G (2004) Deflated citizenship: Labor rights in a global era. In A Brysk and G Shafir (eds) *People out of Place* (pp 109–129). New York: Routledge

Siltanen J (2002) Paradise paved? Reflections on the fate of social citizenship in Canada. *Citizenship Studies* 6(4):395–414

Tabb W (2001) *The Amoral Elephant*. New York: Monthly Review Press

Tieleman B (2002) Business is reborn without labour. *The Georgia Straight* May:23–30

Vosko L (2000) *Temporary Work*. Toronto: University of Toronto Press

Waldinger R and Litcher K (2003) *How the Other Half Works*. Berkeley: University of California Press

White J (1993) *Sisters in Solidarity*. Toronto: Thompson Educational Publishing

Chapter 3

Maria's Burden: Contract Cleaning and the Crisis of Social Reproduction in Post-apartheid South Africa

Andries Bezuidenhout and Khayaat Fakier

Introduction

A day in the life of Maria Dlamini, a 45-year old black woman working for Supercare[1] at the University of the Witwatersrand[2], starts in Soweto at 3:45am. After she has boiled the water for her bath and tea, and had breakfast, it is 4:45am and she heads for the minibus taxi[3] pickup point, a 10-minute walk in the dark from her home. When she gets into the taxi at 4:55am and pays R5,[4] it is already filled with fellow workers and blissfully warm. Soon the taxi is filled to over its capacity by 14 others—mostly women—and reaches the M1 carriageway as everyone naps in their seat. The taxi reaches Braamfontein, downtown Johannesburg, at 5:30am and, battered by the wind, she walks for another 25 minutes to University Corner. She waits until 6:15am, then "clocks" in at the Supercare offices. By 6:40am, she reaches her small room on the third floor of Senate House. This room is warm and narrow, and contains two chairs, a table, Maria's cup and plate, and laminated Wits calendars. She naps again until 7:30am. Although her working day officially starts at 7:00am, Supercare management will not know of this extra half hour of sleep.

For the rest of the working day she will clean over 30 offices on two floors of Senate House. During her lunchtime, she makes her way to ABET[5] classes on West Campus where she learns basic reading and arithmetic skills. Although her day started when most of Wits's staff was still asleep, she is always friendly and well liked by everyone. At 3:40pm she makes her way back to University Corner where she clocks out. She reaches the taxi rank at 4:20pm and 45 minutes later she is dropped off at the end of Myundla Street in Phiri, Soweto. She first buys a half liter of milk, a luxury at R5, then heads to her home about eight houses further up the dirt road.

Maria arrives home at 5:25pm, 12 hours and 40 minutes after she left, tired, hungry and overwhelmed as her day's work is not yet complete. First, she eats the food she cooked on Sunday. If there is food left over from her allotment for the day, she gives it to the 2-year-old twins of her "niece",[6] who calls Maria "Mam'khulu".[7] After supper, she briefly visits her aunt, the grandmother of the twins, a pensioner who has been plagued with severe pains in her arms and legs. By 6:00pm Maria is again hard at work sewing, either doing alterations on the clothes of Wits workers or making her own clothes. Finally, at 8:30pm, she goes to bed, exhausted, in preparation for the cycle repeating itself the next day. This, then, is a typical working day for a female cleaner at Wits. The day is long, interspersed with trying to catch naps. It is not just a day of *production* but also of *reproduction*, with no clear boundaries between the two. In fact, production is found to penetrate what is often seen as the traditional realm of reproduction, the household (Elson 1999; Humphries and Rubery 1984; Peck 1996; Pocock 2003).

Through recounting Maria's story—a story similar to that of thousands of other workers—we explore how the boundaries between paid formal work and the household are changing for a number of outsourced cleaners at Wits. We shed light on the broader transition in the nature of work as outsourcing is increasingly replacing the more traditional standard employment relationship (SER) and argue that the social consequence of this system—that of shifting the burden of social reproduction onto households—replicates one of the key characteristics of apartheid. For sure, the ways in which boundaries of inclusion and exclusion in the formal labor market are now being drawn (Silver 2003:20–25) are no longer determined by the contours of a state-sanctioned regime of apartheid racial classification. Rather, the struggle over "defining the content of working-class 'rights' [and over] the types and numbers of workers with access to those rights" (Silver 2003:21) has become increasingly embedded in the neoliberal language of the market. Yet, as we will show, whilst the drawing of these boundaries now increasingly reflects the micropolitics of institutions such as universities rather than the policymaking of national-level government, and whilst the contours of such boundaries seemingly now turn on apparently neutral determinations such as improving economic "efficiency" rather than on blatantly racist efforts to implement a rigid, racialized labor hierarchy, the nature of these boundaries is certainly not devoid of notions of "race" or, for that matter, of gender. In fact, the legacy of apartheid is clearly fusing with the logic of the market to link the present outcomes and future possibilities for the working poor to the constraints of the past. Indeed, it is proposed here that the discourse of economic efficiency in post-apartheid South African workplaces actually relies upon mechanisms of the apartheid labor regime, thus exacerbating the dire circumstances of thousands of households struggling with poverty, unemployment and HIV/AIDS. The result, we suggest, is that these households, conceived of as "sites of stability" (Mosoetsa 2003) and reproduction, are under considerable threat as managers appear to "be resolving the crisis of the post-apartheid workplace order by displacing confrontation, antagonism and disorder into the family, the household and the community . . . [a solution which] generates broader social crisis whose symptoms are the breakdown of social solidarity, intra-household and community conflict, substance abuse, domestic violence, and the proliferation of other crimes" (Von Holdt and Webster 2005:30).

And yet, it was not supposed to be this way. With the formal ending of apartheid in 1994 and the coming to power of the African National Congress (ANC), many had hoped that the legacy of apartheid labor practices would soon begin to be overcome. Certainly, the ANC's efforts to reform the regulatory and institutional labor relations regime have often been presented as an example of good practice, and the South African case has featured prominently in debates in the International Labour Organization (ILO) about the challenges posed by new forms of employment, such as triangular employment relationships and what the ILO calls "disguised employment" (Cheadle and Clarke 2000; Clarke, Godfrey and Theron 2003). However, as we will show, the context of neoliberalism within which reform is being undertaken is having a substantial impact upon the unfolding of new employment relationships and the spatialities thereof (Herod 2001). Hence, our analysis shows that the seemingly progressive institutional arrangements explored in the debates over employment relations reform do not appear to be capable of improving the lives of ordinary workers in a post-apartheid, though increasingly neoliberal, South Africa. Indeed, as we outline below, it is clearly the case that the way in which contract cleaning at a place like Wits now operates does not differ significantly from pretty much anywhere else in our neoliberalizing world (see, eg, Bernstein 1986; Puech 2004; Zlolniski 2003).

Our study, then, confirms the findings of a number of others which have shown that the protective elements for workers of a labor regulatory regime which is itself predicated on long-term, stable employment relations between singular employers and employees within a system of industry-wide collective bargaining arrangements do not appear adequately to account for the nascent complex triangular employment relationships[8] which are characteristic of contemporary neoliberal labor practices (see Kenny 2003; Kenny and Bezuidenhout 1999; Theron 2003; Theron et al 2004; Von Holdt and Webster 2005). What makes this all the more significant in the field of higher education is the fact that higher education is subsidized by the state and is thus not exposed to "global competition" in the way that other sectors of the economy are, such as mining and manufacturing. At least in theory, then, of all sectors it is the educational sector which should be able to provide for stable employment conditions, although in practice this does not seem to be the case.[9]

The analysis here draws upon a number of data sources[10]—interviews, observation and existing documents. In examining the impact of outsourcing, we begin by presenting an overview of the apartheid labor regime, focusing on how the burden of reproduction was systematically shifted onto black households. We then describe how resistance to this system led to its reform. Amid these attempts at reform, the re-segmentation of the labor market under neoliberalism has reintroduced a logic that is, paradoxically, similar to that of the apartheid labor regime. We subsequently look at the dynamics of workplace control as employment relations at Wits have been exposed to outsourcing. Finally, we explore the impact of contract cleaning on households.

The Rise and Decline of the Apartheid Labor Regime

Whilst not exceptional when compared with other southern African colonies, the apartheid system in South Africa was, however, extreme in its racial ordering of society. According to Wallerstein, Martin and Vieira (1992:4), the colonization of southern Africa was peculiar for two reasons. First, the region's "lands were the site of incredible mineral wealth"; second, the region had "a white settler population of some size and cultural cohesion who were determined to maintain and enhance their group power and status". This presented a dilemma, for whilst the mining industry required cheap labor, the settler population generally saw the urbanization of the African population as a social and political threat. To solve this dilemma, the migrant labor system was introduced. Consequently, labor was secured through the destablization of rural economies undertaken via the linked measures of legal dispossession of land and the introduction of various taxes—to be paid in cash—which required rural dwellers to work for wages. As a way of controlling such labor, African men who migrated to urban areas and the mines were typically housed in single-sex hostels, whilst their wives remained in rural areas to raise children and take care of the aged through engaging in subsistence agriculture and using remittances sent to them by their husbands. For their part, mineworkers were employed on fixed-term contracts and had to return to their rural homesteads when those contracts expired or when they became ill (Moodie 1994). Thus, "grand apartheid"[11] ensured that the mining industry could subsidize its profits by externalizing a significant part of the burden of social reproduction to rural subsistence economies (Wolpe 1972). In the case of Johannesburg, founded in 1886, the influx of labor associated with the Witwatersrand gold rush soon resulted in the city becoming South Africa's most densely populated metropolis. For its part, Wits University was initially established in Johannesburg as a training school for the mining industry. Academic members of staff were predominantly white and male, whereas support staff members were African (Murray 1982).

Apartheid's politics of boundary drawing was enshrined in one of the key legal pillars of the system's labor regime, namely the Industrial Conciliation Act of 1924. Central to this system of labor control was the concept of the "pass", a document which dictated where its bearer could and could not be at any particular time. Significantly, the 1924 law excluded "pass-bearing natives" from the definition of employee. Initially, this applied solely to African men, since pass laws were applied to African women only from the 1950s onwards. Later on, the state would also come to enforce the racial segregation of trade unions, whilst anti-communist laws would be used to neutralize labor activists (Baskin 1991). Broadly speaking, though, the apartheid labor regime had three key characteristics.

First, it was based on a racial division of labor. Color bars were enforced with regard to both jobs and wages (Johnstone 1976) and a racially segmented labor market was developed, with black workers progressively occupying the positions of laborers, assistants, and later "semi-skilled operators" (Webster 1985), whilst whites were artisans and managers. Wits University reflected this racial division of labor. Whilst the professoriat was historically white, Africans were employed to do manual work in support functions and were mostly supervised by white employees (Murray 1982; NEHAWU 1985; Shear 1996).

Second, the urban geographies created by the migrant labor system served as an additional form of labor control. Cities were divided into white "suburbs" and black "townships", with industrial areas as "buffer zones". African women from townships commuted to suburbs to clean houses, rear children, and cook food for white families. They were only allowed to do so if they possessed passbooks, signed by their employers (Cock 1980). In terms of grand apartheid, rural labor reserves (called "homelands" or, less politely, "bantustans") became "self-governing states" where Africans were granted pseudo-citizenship rights. In an attempt to stem African urbanization, subsidies were granted to firms setting up industrial activities in such homelands (Hart 2002). This migrant labor system was at the center of the apartheid economy, and particular notions of racial and ethnic citizenship were imposed within this spatial order. Hence, as grand apartheid played out, the goal was seen to be ultimately to declare the African homelands independent states, with the result that their black "citizens" would effectively become foreigners in "white South Africa".[12] Within this set-up, the pass system controlled the movement of labor whilst within urban areas strict racial categories were used to enforce residential segregation. Under this system, workers who lived in company or municipal hostels were in subordinate positions to those who lived in urban townships; for hostel-dwellers, the loss of one's job meant losing one's accommodation and a forced return to the rural areas from whence workers had come (Von Holdt 2003). In the case of Wits, the university maintained an on-site hostel for migrant workers up until the mid-1980s (NEHAWU 1985).

Thirdly, the government formally legislated the racial segregation of facilities, such that firms had to provide separate canteens, change houses and toilets for white and black workers. Although this legislation was repealed in 1983, as Von Holdt (2003:29–30) points out, many firms nevertheless continued with their previous practices well after the law's abolition. For its part, Wits formally maintained segregated facilities for white and black support staff until at least 1985 (NEHAWU 1985).

Despite efforts to maintain the practices and spaces of apartheid, though, South Africa's racial order could not stem the rise of militant trade unions from the 1970s onwards when workplaces were increasingly confronted by rapid industrialization and changing labor processes. As trade unions realized that they would be unable to change the workplace order without removing apartheid, a form of militant social movement unionism emerged in the manufacturing and, particularly, mining industries (Crankshaw 1997; Seidman 1994; Webster 1985). With the decline of the apartheid state, headway was also made in the public sector. Thus, the National Education, Health and Allied Workers' Union (NEHAWU) was formed in 1987, with Bheki Mkhize from the Wits University branch elected as the union's first president (Molete and Hurt 1997:11). The first recognition agreement between this union and any university was signed at Wits, in 1989 (Molete and Hurt 1997:62).

After South Africa's first democratic election in 1994, the subsequent process of political democratization opened up space for the labor movement to influence policy-making even further. In this regard, the Labour Relations Act (LRA) of 1995, a key piece of new legislation in the arena of the workplace, was passed after intense negotiations and delicate compromises. The approach adopted under the new legislation rested on the assumption that institutionalized collective bargaining would move workplace relations beyond the "adversarialism" of the apartheid era. Consequently, the LRA maintained and strengthened the system of workplace collective bargaining. Industrial Councils were renamed Bargaining Councils and were allowed to be set up in industries with sufficient union and employer representation, whilst Statutory Councils were established to facilitate a process of centralization in collective bargaining in industries where no Bargaining Councils existed. The civil service and the agricultural sector, for instance, were brought into the labor regime through the establishment of Bargaining Councils to structure wage negotiations for these sectors.

In addition to drafting legislation that attempted to reconstruct the judicial architecture for collective bargaining on a non-racial basis, legislation was drafted to put in place minimum standards (the Basic Conditions of Employment Act of 1998), to promote employment equity (Employment Equity Act of 1999), and to overhaul the system of vocational education and training (eg the Skills Development Act of 1999). Hence, the stage was set for workers to use a range of new and reformed institutions to unmake the legacy of the apartheid labor market. However, as we now show, there were a plethora of new strategies devised by employers to prevent workers from asserting their newfound rights.

Re-segmentation and the Post-apartheid Labor Regime

Whilst the struggle against apartheid led to democracy only in 1994, labor reforms were in fact being implemented from the late 1970s onwards. Nevertheless, with the rise of non-standard contracts of employment in the mid-1990s—contracts typical of the "new", neoliberal economy—there was a significant re-segmentation of the labor market (Bezuidenhout et al 2004; Kenny and Webster 1999). Alongside this, South Africa's unemployment rate increased from almost 16% in 1990 to more than 30% in 2002. Jobs classified as being in the informal sector grew as a proportion of all employment, from 1.7 million in 1990 to 3.5 million in 2002 (UNDP 2004:238–239). This structural shift in the labor market reduced the ability of many households to withstand the shocks of economic restructuring. Simultaneously, the state pursued the privatization of services such as water and electricity, adding to the cost burden of poor households. Indeed, partly due to the devastating impact of the HIV/AIDS pandemic, South Africa's Human Development Index (HDI) declined from a high point of 0.735 in 1995 to 0.666 in 2002, a figure which equals the levels of the 1970s (UNDP 2003).

In this context, the implementation of neoliberal policies in the South African higher education sector led to severe cuts in state support for universities, inducing a cost-recovery approach that has been described as the "marketization" of higher education. Within universities, power gradually shifted from academics to full-time managers, often resulting in a crude managerial discourse used to express institutional imperatives. Less "profitable" departments were rationalized or closed down, whilst support services were outsourced to private sector providers (Bertelsen 1998). These trends are reflected in the increase in the number of executive and managerial staff employed by South African universities, from 1125 in 1994 to 1229 in 1999. Administrative staff also increased, from 9769 in 1994 to 11,750 in 1999. Over the same period, the number of support staff declined from 14,346 to 10,817 whilst those in trades or crafts dropped from 1433 to 951 (Van der Walt et al 2003:277).

In an effort to mitigate the rapid reduction in their membership brought about by such changes, the NEHAWU went on the offensive by attempting to use the reformed labor regime to stabilize collective bargaining in higher education. By 1997, the union was recognized at 18 of South Africa's 19 universities and at 12 of the 15 technikons (polytechnics) (Molete and Hurt 1997:63). In February 1999, the union made clear its intention to set up a Bargaining Council for the higher education sector[13] by serving notice to the National Economic Development and Labour Council (NEDLAC) under the terms of Section 77 of the LRA, which allowed workers to embark on legal industrial action to promote their "social and economic interests". NEDLAC then arranged a meeting for the union with the South African Universities Vice-Chancellors Association, the Council for Technikon Principals and the Department of Education. Subsequently, in August 1999, a summit was held which set up a national forum to take forward issues such as "the future of the collective bargaining relationship", "employment equity requirements", and "financing of tertiary education" in the sector. Although the union was disappointed by the outcome of the summit, pending these further consultations, it agreed not to engage in industrial action to pursue its goals.

In response, however, universities' managements began an unprecedented and seemingly coordinated attack on the union membership from "liberal" universities (such as Wits). As part of this, in 1999 Wits engaged a management consultant firm—University Management Associates (UMA)—to investigate the restructuring of support services. A laborious process of "consultation" with various staff representatives was staged, since consultation when retrenchment was considered was a requirement of the Labour Relations Act. It was almost inevitable, however, that this process would lead to a decision to outsource most support services, including cleaning (see Adler et al 2000). The consultants had used the same process at the Universities of Cape Town and Pretoria, and so, perhaps not surprisingly, the exercise at Wits yielded the same result—outsourcing. In all three cases, the decision ended up in the Labour Court, and in the case of the University of Cape Town, in the Constitutional Court.

During the debates about how to restructure higher education, outsourcing was constantly presented as a more affordable option for university administrators, based on the assumption that contract cleaning firms had developed sophisticated cleaning capabilities based on specialized equipment and appropriately trained workers. Thus, the National Contract Cleaners Association (NCCA) has argued, in one of its promotional pieces on the website of the *Proudly South African* campaign (a campaign initially funded in part by the government's Department of Trade and Industry), that the world has changed since the days when Wits's employment practices—including the use of in-house cleaning staff—were first established: "[I]nternationally there has been a marked shift to outsourcing and using the services of contract cleaning companies as more businesses are keen to get on with their core activities". If South African employers want to keep up with global trends, the NCCA has maintained, they will have to consider why contracting out cleaning is such an attractive option and what advantages it may provide in the area of "labour disputes, recruitment and personnel costs . . . [and in the] time needed to oversee and administer cleaning processes". Furthermore, NCCA has suggested, unlike the case of in-house cleaning operations, professional cleaning companies "have invested in skilled and efficient staff, the correct equipment and [through] the purchase of materials in bulk . . . can keep costs to a minimum", whilst outsourcing protects the client against third party liability.[14]

In the Wits context, UMA consultants argued that outsourcing would provide the "most significant benefits in terms of cost, productivity and efficiency of operations, possibly resulting in substantial improvements in service levels and the cost of delivery".[15] Even contract workers, they claimed, "could benefit from greater career opportunities, development/training and accreditation".[16] The image created, then, was that of a modernized, specialized cleaning service which would allow the University to focus on its core functions (teaching and research). Consequently, on 25 February 2000 the University's Council passed a resolution to ratify an earlier decision to outsource five major support services. University officials claimed that they could save costs by increasing efficiency and specialization. However, it soon became obvious that outsourcing was also designed to undermine unionism on university campuses. Thus, a draft document handed out to prospective contractors contained a clause that required contractors to "discourage" their employees "from participating in any industrial action". Moreover, if these employees were "*guilty* of participating in industrial action", the university could require contractors to "remove them from the client's premises".[17] Through such a clause, then, "industrial action", which is protected by the South African constitution, would be criminalized.

Despite protests from students, workers and some faculty, Wits pressed ahead with the outsourcing process, awarding a contract to

Supercare. As a result, workers' wages were slashed to a third of previous levels and they lost most non-wage benefits, such as medical aid and financial aid for their children to study at the university. At the same time, the local branch of the union was all but closed down. Clearly, although outsourcing may have provided advantages to the university administrators, its consequences for workers were devastating and would lead both to a dramatic intensification of work in the workplace itself and, as a result of the reduction in wages and benefits, to the shifting of the burden of social reproduction to households and communities in the urban areas.

Contracting, Control and Work Intensification

UMA consultants presented outsourced cleaning as being more cost-efficient because of the use of specialized equipment and efficient management. In contrast, our research shows that costs have been cut primarily through work intensification and a reduction in wages and conditions. Indeed, cleaners at Wits are now subject to a whole range of pressures to which they previously were not. Primarily, they have to complete an amount of work which a workforce double the present size could most likely not do efficiently. Thus, whereas before restructuring Wits employed approximately 600 cleaners, Supercare now employs only 280 workers, most of whom previously had worked directly for the University and who had been retrenched when cleaning services were contracted out (Van der Walt et al 2003).[18] This means that each cleaner has approximately 30 offices to clean between 7am and 3:30pm. At the same time, they battle with cleaning materials and equipment of a low quality, they must monitor the quality and extent of other workers' performance, they are often summarily accused of theft if there is no other explanation for missing supplies, and they must carry out the work of absent or negligent fellow workers. All of these pressures occur in an environment of excessive control, which has resulted from the inability of Supercare's management to supervise the work process directly because of their physical distance from the workplace; the geography of the Wits campus—spread over almost 100 ha—means that Supercare cleaners work in widely divergent parts of the University, making direct control and supervision of these workers difficult for the Supercare management, who occupy offices about 200 m outside the campus boundaries.

As a consequence of this inability to engage in much direct control of the workforce, Supercare management exercises its rule through four forms, all of which represent a worsening of cleaners' conditions of work relative to previously: firstly, through intensified supervision; secondly, through management exhorting the workers to watch each other; thirdly, by encouraging Wits employees to serve as "adjunct managers"; and finally, there is "control through abuse" on the part of the Supercare management.

For supervisors, the primary activity in which they engage is to ensure that workloads are carried out. Certainly, some are sensitive to the plight of the cleaners. Thus, one supervisor we interviewed [Elizabeth] was genuinely liked by the cleaners because she was seen to be sympathetic to their workload: as one reported, "Elizabeth takes work from my hands when it is too much".[19] As Elizabeth herself recognized, the cleaners "have too much to do . . . [There is] one person to clean over 30 offices and two people to clean 14 lecture theatres during the day when school is on. How can I scold them for not doing the work when the time is not enough? I am running between the [Supercare] office and Central Block [the building she supervises]".[20] However, this sympathetic attitude does little to ease the cleaners' excessive workloads, whilst the pressure on Elizabeth herself to ensure the work is completed has led her to be constantly on the lookout for other employment.[21] Furthermore, such an attitude amongst supervisors is rare and is, perhaps, partly due to the fact that Elizabeth is one of only a few African supervisors. Indeed, the racial segmentation of the internal labor market has been carried over from the apartheid workplace, represented by the fact that Supercare mostly employs "colored" women in supervisory positions, many of whom have little empathy with the African cleaners—itself largely a holdover from the former's (relatively) privileged positions in the racial hierarchy of the apartheid era.[22]

With regard to the cleaners themselves, Supercare has sought to exercise control on the expansive campus through turning workers into extensions of management's gaze. Thus, in response to the theft of toilet rolls and cleaning materials,[23] Supercare management began to pressure workers to watch each other. This was accompanied by separating the task of dispensing supplies from the actual cleaning of toilets and offices. In this way, the dispenser watches the cleaner for excessive loss or use of these consumables and vice versa, and control is accomplished by the workers monitoring one another. For their part, whilst Wits staff are largely ambivalent about the quality of service provided by Supercare, some do recognize that caution must be taken in complaining to the company, as some staff are concerned about the repercussions for Supercare workers. This concern is not baseless, as the Supercare area manager confirms that, no matter how much the work of a cleaner is valued, if two complaints are made against the cleaner she will be removed from that work area.[24] This arbitrary exercise of power (which was not a feature of the brief period when cleaning was done in-house in the post-apartheid era), however, returns to a key aspect of the apartheid workplace regime, allowing

Wits employees—should they choose to do so—to capriciously punish cleaners for whatever reason they deem appropriate. "Control by abuse" is also quite common. Thus, one of us observed that after Supercare management received complaints about dirty offices and toilets, the area manager walked to the worksite concerned and started shouting and swearing at the worker in full view of passers-by, using terms like "*kaffir*" and "*swart bliksem*".[25] That this is not uncommon was reported in the *Wits Student*, a student newspaper, which cited a case of physical and mental persecution of a pregnant worker that led to the miscarriage of her child (Ngcobo 2004). Such abuse appears a necessary constituent of control, which sustains the apartheid racial structure of power into the post-apartheid era.[26]

Although all the women cleaners recognized Supercare as their employer, significantly many closely identify with Wits University staff members. For instance, we observed that Maria had only two photographs of herself with kin yet displayed five photographs of herself with Wits staff, whilst her bedroom was decorated with gifts from Wits staff. The walls of her living room were decorated with laminated Wits calendars and signboards. Likewise, whilst the cleaners' leave arrangements are the responsibility of Supercare management, the cleaners would often inform Wits staff of their leave arrangements to prepare them for the arrival of replacement staff or to seek their advice.[27] These identifications with Wits are key for the ability of both the University and Supercare to construct an external-ized relationship between the cleaners, Supercare and Wits. Hence, whilst there is a clear recognition by all concerned that Supercare is the direct employer of these women, our research indicates that Supercare is viewed merely as a "nominal" employer (Theron et al 2004:311) and it is Wits which provides the worksite and with whom the workers identify and therefore see as their "real" employer. Legally, the relationship between Wits management and the cleaners has been transformed from an employment relationship to a commer-cial relationship between Wits and Supercare.[28] In the process of externalization, the material claims that workers can make on Wits have been reduced, whilst the degree of control over workers has been enlarged with the introduction of the nominal employer. Nonetheless, in their own conceptions of their work many cleaners appear to see themselves as still "working for Wits", an understanding which has interesting implications for the construction of their own identities as workers and so, perhaps, for their views on how to go about resisting such changed working conditions.

This ambiguity is significant, for both Supercare management and Wits administrators conspire to exploit the legally clear, but emotively nebulous, position within which the cleaners find themselves. For instance, when we interviewed the Human Resources Director at

Wits and asked what leverage Wits has to change the conditions of the cleaners' employment and, indeed, to improve the cleaning service provided by Supercare, he stated that Wits chose not to use this leverage, as it would have cost implications—with every condition attached to the contract between Wits and Supercare, extra costs are included.[29] Whilst Supercare and Wits rely on externalization to avoid responsibility for the conditions of work at the University, they simultaneously collude with each other to prevent the cleaners from contesting their conditions. Thus, Supercare workers have tried to establish a trade union, yet Wits uses its legal ownership of the campus to prevent the cleaners from using any of its buildings for mass meetings. At the same time, although Supercare cannot deny the workers the right to meet—which is protected by the LRA—their externalized relationship allows them to determine that union meetings must be held at Supercare premises, approximately 25 km from Wits. The enduring apartheid geography of South Africa—reflected in the vast distances cleaners would have to travel from their homes in Soweto or from the university campus in order to participate in a union meeting in the white neighborhood of Randburg in northeast Johannesburg in which these premises are located—thus renders mass meetings unaffordable and impossible.

As we have seen in the discussion above, then, despite the claims of neoliberals, outsourced services are often inefficient and involve the hidden costs of managing complex relationships, lowered productivity and declining morale of the workforce. These costs accrue to both the client and the contractor. As we shall see in the next section, however, the costs to workers are often hidden, or ignored, and transferred to households and communities. In the context of a neoliberalizing South Africa, the gendered and racialized nature of outsourced work, we argue below, means that this transference adds greatly specifically to the burden of those who have long been marginalized by apartheid— the female, the African, the unskilled and the aging.[30]

Work, Households and Social Reproduction

The inability of Supercare workers to organize themselves into a union means that they cannot use institutional mechanisms to contest their wages and working conditions. These are set by a Sectoral Determination in terms of the Basic Conditions of Employment Act (Republic of South Africa 2004). Before the outsourcing process took place, Wits cleaners earned more than R2500 per month. They were part of a provident fund, members of the university's medical aid scheme, and their children could receive scholarships to study at the university. After their jobs were outsourced, their new status as non-core workers left them with significantly reduced benefits. For example, those interviewed for this study earned between R500 and R1230 per month. Thus, Susan and Nozipho both worked on short-term contracts for Supercare and earned R500 and R797 a month, respectively. Lindiwe, who had worked at Wits for 14 years, earned R1000 a month, as did Joanne, who had worked at Wits for eight years. Whilst Supercare does have a provident fund, two significant benefits have been taken away—membership of a medical aid scheme and scholarships for cleaners' children to study at Wits. At the same time, cleaners' leave days have been reduced from 24 days a year to 12 days.[31]

As seen in the case of Maria, a Supercare cleaner's workday generally starts long before 7am, the official starting time. Women cleaners report their time of waking in the morning as ranging between 3am and 5am and they arrive back home as late as 6:30pm. The cost for transport varies between R100 and R300 per month, up to 30% of the workers' incomes. Special arrangements have been made for a "Wits" taxi to pick up workers in Phiri, Soweto—a "town-ship" designed during apartheid for the residence of African workers servicing the needs of "white" Johannesburg—but, significantly, this is by agreement between commuters and not with Wits. Therefore, three to four hours of any working day comprise simply getting to and from work. This is a problem of particular significance for black South African commuters who still reside in geographically separate townships, quite far from their workplaces. The lack of adequate public transport and the long distances workers have to walk to their pick-up points render the simple act of going to work both expensive and, often, unsafe. Hence, whilst Maria contended that all the walking kept her young and fit, the fact that her (asthmatic) chest would start to wheeze at 6pm is testament to the effect of traveling and walking unprotected in the extreme cold of the early morning.[32]

The working day of a cleaner does not end, however, when she gets home. Because she is a woman, most of the household tasks remain her responsibility.[33] Only two of the ten women we interviewed have help with taking care of the home. The rest are primarily responsible for cooking, cleaning and the maintenance of the home. Significantly, none of these women rely on their spouses or boyfriends for assistance at home. Equally, hiring domestic care is not an option. Instead, this burden is added to their long list of daily responsibilities. Indeed, the workload for these women is quite significant and is exacerbated by the fact of their low wages.[34] Hence, in order to stretch her money, rather than purchasing pre-prepared food from the supermarket, Maria would buy two live chickens every two months, slaughtering, cleaning and cutting them up herself. In addition, she would bake her own bread and cook supper for the whole week on a Sunday afternoon.[35] In order for poor women to ensure the survival of their families on meager incomes, then, they will subsidize the budget for food with their own labor (Pearson 2000). Thus, they buy cheaper cuts of meat that require longer cooking and grow their own vegetables, which decreases the need to spend money but significantly increases the time women spend on domestic care.

Moreover, only three of the women had partners or children who contributed to the household income. Any money entering the household is spent on rent, food, electricity, transport and children's school fees. The lack of support in their homes and their low wages deprive these women of the possibility of hiring services, which would lighten the load of reproduction though would, arguably, merely pass off onto other similarly situated women such domestic chores.[36] To top this off, cleaners' income from sources other than Supercare invariably exceeds their salaries. Nevertheless, this other income is still too small for cleaners to give up their jobs with Supercare. Thus, in addition to sewing, Maria turned two rooms in her backyard into rented accommodation for lodgers. Transforming her home into a means of generating income was a relatively easy way of making extra money, as opposed to her more labor-intensive sewing.

An in-depth look at the conditions of Maria's work and household, then, points very clearly towards the intrusion of production into the sphere of reproduction. This has significant implications for such workers. There is evidence, firstly, of how the intensified nature of work affects cleaners' health in a negative manner. Chemical cleaning products are used by Supercare regardless of cleaners' complaints of their effects on their health, and no masks or gloves are provided either.[37] For Maria and Joanne, this was especially hazardous as they suffered, respectively, from chronic asthma and high blood pressure and could not afford to visit a doctor, as Supercare provides no medical aid benefits in the way in which Wits did in the past.[38] Secondly, because of the lack of benefits—in the form of medical aid, housing subsidy and pension—cleaners have to generate additional income in order to meet their financial needs. The additional pressures of the lack of adequate public transport and health care, as well as excessive charges for water and electricity provision, aggravate their financial situation to the point where they are slowly eroding their savings. This means that, in addition to her formal employment by Supercare at Wits University, Maria's productive, income-generating activities were extended to her home, in the form of sewing and providing rooms to lodgers.

Maria's emotional attachment to Wits was exhibited by her photographs of the University and staff members, but she had another link at home through her sewing. In addition to providing her with a place of work, she was also able to access a market for her sewing skills. As her family and neighbors were not able to pay her for sewing, she became dependent on Wits staff to support this activity.[39] Maria, though, was not alone in facing such hardships. Seven of the women, all unmarried, were engaged in some form of additional income generation and subsistence activities, ranging from the growing of vegetables for their own consumption to renting out rooms. Given that their wages are now about one-third less than when they worked directly for Wits, such activities constitute an ever-more significant portion of their overall income. Sometimes these women would exchange the vegetables they grew for other foodstuff with neighbors. Although two of these women had boyfriends, these men did not contribute to the household income nor take part in income generation activities.[40]

Six of the women interviewed had small children and relied on help with childcare from their mothers, in-laws or sisters. The men in their lives—husbands or boyfriends—did not assist or take responsibility. Significantly, three of these women sent their children to reside with their families in rural areas, whilst Lindiwe's mother, who lived nearby, took care of her sickly child. This is quite typical in South Africa, as childcare, especially of smaller children, is integrally linked to extended networks of family and kin (Mosoetsa 2003). These networks extend deep into the rural areas of South Africa, a fact which is itself a legacy of the apartheid labor migration regime when "non-productive" older women and children were spatially restricted to rural homelands through denial of a pass. However, "outsourcing" one's childcare to the rural areas is often no less costly than are formal paid childcare arrangements, for although the rural caretakers of children are not always paid with money they nevertheless usually must be compensated for their effort through some form of non-monetary rewards. In addition, the women are still responsible for the financial care of the children, even though they do not reside with them. Furthermore, such women are expected to send remittances to their families' rural households, not only for the maintenance of the children sent to live there but for the survival of the rural homestead itself. This all places a great burden on the cleaners, although the rural homestead nevertheless does provide a site of support for social and biological reproduction, reproduction which is not adequately supported by urban life.

Although an extended network of kin supports the care activities of the household and indirectly supports women's endeavors in the workplace, the income of any one home also supports the homes of an extended network, which does not necessarily comprise only kin. Hence, Maria was the locus of a very intricate web of relations defining her social and economic relations with family members and other dependants. Previously, Maria had had a relationship with a migrant worker who worked in the goldmines of Gauteng. When he and, later, his wife died, Maria adopted their two children. She provided support for other households as well. A hundred meters from her home lived her aunt, who had two daughters—one at school and the other unemployed and the mother of twins. Although these children were primarily supported by the aunt's sons, Maria provided crucial support by sharing her food with the twins and their mother. She also supported her unemployed sister and brother-in-law and their nine children in QwaQwa, in exchange for their providing a home for Maria's two adopted children. Her brother-in-law's brother, who resided with the family, also supported this household. Maria's other sister, who lived and died in Zimbabwe, was survived by two unemployed children, who were supported by Maria and their three brothers.

Although Maria was not the primary provider of any of these 21 members of her extended network of relations she would buy their groceries, shoes and underwear and make their clothes along with her own. Visiting their homes not only involved transport costs but also the expense of "gifts" comprising necessities such as staple foods without which these households could not survive and which she managed to buy through conscientious saving. It is inconceivable that these far-flung nodes would have been able to survive without her contribution, a fact which supports Mosoetsa's (2003) contention that poor households depend greatly on the support of their extended networks of relations, as their own impoverished communities can offer no relief. Despite the austerity of her life, though, Maria's daily toil was given great meaning by the love and emotional support she received from her "family". Maria never married, nor had children herself, but she was able to state that: "Everyone loves me, they don't give me money, but they love me".[41]

Conclusion

The transition from apartheid to a democratizing society has been described as a triple transition. This implies simultaneous but often contradictory processes of political democratization, economic liberalization and the deracialization of society. Thus, the many dimensions to the South African transition result in a tension between political, economic and social forces (Von Holdt 2003; Webster and Omar 2003). Whilst this understanding is useful in explaining broader changes, it tells us little about the micro-politics of transition.

Our discussion of the changes in the internal labor market of Wits provides us with a vantage point from which to explore these micro-processes. It enables us to make sense of the changing contours of the new order and to explore continuities and discontinuities with the past. Grand apartheid's migrant labor system and government attempts to impose a national political geography based on "race"—a "white South Africa" separate from the "black homelands"—led to the burden of social reproduction being shifted onto households and communities in the rural areas. We argue that the system of outsourcing maintains this logic, and that this is compounded by the rise in unemployment and the privatization of services. Thus, in the case of Wits, although the University has cut institutional costs by contracting out cleaning, it has done so at the expense of those workers at the bottom of the socio-economic hierarchy, with the result that, as was the case under apartheid's migrant labor system, the bulk of the cost of social reproduction is increasingly being carried by rural and urban households, and not the primary employer or the state.

Furthermore, in the context of post-apartheid South Africa, neoliberal policies are undermining attempts to deracialize the workplace. As our research shows, there is a need to investigate how boundaries of exclusion and inclusion are being reworked. These boundaries are no longer determined by the crude racial language of apartheid. Instead, they are expressed through struggles over the nature of the employment relationship and the responsibilities of the employer. Who will benefit from the new order is determined by seemingly innocent questions: "What are the 'core functions' of an institution?" and "Who forms part of the university as a community?" Indeed, these struggles call into question the very definition of a work*place*. Moreover, it is perhaps no accident that contract cleaning reintroduces many of the aspects that characterized the apartheid workplace regime, such as job insecurity, low pay, and a lack of benefits. In a nutshell, cheap labor is being reproduced under new conditions as the burden of reproduction shifts onto households and communities in both urban and rural areas. Exclusion is no longer expressed as a racial imperative but as a force necessary for the efficient running of public institutions. This form of social exclusion leads to the fragmentation of workers into a core of stable, full-time employees and a periphery of insecure, externalized workers, a distinction based on the discourses of market.

The working life of Maria Dlamini illustrates the realities of South Africa's transition, as neoliberal policies are implemented. Sadly, shortly after we completed our research, Maria fell seriously ill. She had been struggling for some years with asthma and was on unpaid leave, when, in October 2004, she died. Her death certificate states that she died of "natural causes", but had she had adequate access to health services her death could probably have been prevented. Her colleagues had to contribute their meager resources to ensure she had a dignified burial. The life and death of Maria, then, vividly illustrate the new conditions under which cheap labor-power (Wolpe 1972) is being reproduced in post-apartheid South Africa.

Acknowledgments

Sections 4 and 5 rely on the fieldwork of Khayaat Fakier and Xolile Mhlongo, conducted for a project titled: " 'Supercare' or 'Care-less'? A case study of the work and living conditions of female outsourced workers at the University of the Witwatersrand". We would like to thank Professors Edward Webster, Andrew Herod, Luis Aguiar, and Chris Benner, as well as two anonymous referees, for commenting on earlier drafts of the paper.

Endnotes

[1] Supercare is a contract cleaning company responsible for the bulk of cleaning at the University of the Witwatersrand. It was renamed Supercare Fidelity after it was bought out by Fidelity and is now listed on the Johannesburg Stock Exchange. For the sake of brevity the name Supercare will be used.

[2] The University of the Witwatersrand is the premier university in Johannesburg, the center of South Africa's mining metropolis. It will be referred to as Wits, the name by which it is most commonly known.

[3] Minibus taxis are privately operated and travel along fixed routes for a pre-determined fee. It is generally the most affordable mode of urban transport in South Africa, although unsafe because of reckless driving and occasional violent conflict between rival taxi associations.

[4] At the time of writing, US$1 is roughly equivalent to R6.50.

[5] Adult Basic Education and Training is sponsored by Wits for Supercare workers who have not completed primary education.

[6] Her niece is actually the daughter of her mother's cousin who, with her mother, sister and twin children, lives a few houses away from Maria.

[7] Literally, "mother big", or "big mother".

[8] Triangular employment relationships refer to cases where a third party enters the picture as a labor market intermediary. In some cases, two or more intermediaries are positioned between workers and the "client". This latter has been referred to as "cascading subcontracting" (see Bezuidenhout et al 2004).

[9] Wits is a particularly compelling case study for two principal reasons: the University was well known during the apartheid era as a bastion of liberalism, and the branch of the National Education and Allied Workers' Union (NEHAWU) at the university was one of the more prominent union branches affiliated to the Congress of South African Trade Unions (COSATU). Indeed, two presidents of NEHAWU, as well as a number of politicians who were active in the ANC, rose through the ranks of the union branch at Wits (see Molete and Hurt 1997).

[10] We draw on in-depth interviews with 10 female contract cleaners employed by Supercare at Wits University and an interview with the university's Director of Human Resources. In order to experience first hand what Supercare workers "actually are doing" (Burawoy 1998:5), instead of relying on information that might have been colored by normative prescriptions, we used participant observation in the home of Maria Dlamini. Our employment at Wits enabled us to use various degrees of participant and non-participant observation in the workplace. We also used documents such as newspaper and union reports to supplement the analysis. All names—with the exception of Maria's—are pseudonyms.

[11] As opposed to "petty apartheid", which refers to the racial segregation of facilities such as park benches or drinking fountains.

[12] In the late 1800s, after the discovery of gold on the Witwatersrand, the Cape Colony legislature passed the Glen Grey Act (1894), laying the major legal foundations for

apartheid's spatial pattern. Eventually, four homelands—Transkei, Ciskei, Venda, and Bophuthatswana—were declared "independent", though only the South African government acknowledged them as such. This meant that "citizens" of these "countries" became foreigners in South Africa and could only travel outside these homelands if they were able to present pass books to the South African police.

[13] Establishing a Bargaining Council—designed to facilitate centralized collective bargain for a sector—was first raised in 1995 in the National Commission on Higher Education. A Bargaining Council in the sector would imply that annual wage negotiations would include all institutions, instead of each university or technikon negotiating wages with the union at a decentralized level.

[14] http://www.proudlysa.co.za//level2/results.asp?include=/docs/news/news/2003/0827.html (last accessed 21 October 2005).

[15] University of the Witwatersrand, Support Services Review, 1999.

[16] University of the Witwatersrand, Support Services Review, 1999.

[17] Taken from an undated and anonymous University of the Witwatersrand document entitled *Facilities Maintenance Management Services* which was handed to all prospective contractors. See also Adler, Bezuidenhout and Omar (2000).

[18] Observation, Supercare, 13 May 2004.

[19] Interview with Phumzile, 5 April 2004.

[20] Interview with Elizabeth, 5 April 2004.

[21] In January 2005, Elizabeth started a new job as a housekeeper at one of Wits's student residences. This is one of the positions that was cut in 2000. However, the university was forced to reinstate such posts as student residences ceased to function effectively.

[22] In South Africa, a distinction is drawn between African (indigenous) people and "coloreds" (mixed race). The Population Registration Act built on popular racial classifications and formalized these into rigid racial categories. Historically, labor market segmentation followed these categories, with "coloreds" in more skilled categories than Africans (see Webster 1985).

[23] Observation, Supercare, 13 May 2004.

[24] Observation, Supercare, 13 May 2004.

[25] Literally translates to "rascal". When the cleaner concerned asked the manager the meaning of this word, she responded that it was the name of her dog.

[26] Whilst the staff of Supercare is predominantly female, all cleaners and most supervisors are "black". However, it is important to recognize that, according to South Africa's warped racial discourse, the term "black" usually includes Africans, coloreds, and Indians. Thus, the cleaners are, in fact, mostly African whereas the supervisors are mostly colored, a line of racial demarcation and hierarchy which mirrors that of the apartheid era. Virtually all the managers above the supervisory level are white. The "site manager" responsible for the Wits contract spanning the whole university was a white male (interview with Wits HR Director, 18 April 2004). Thus, specifically in the instance of outsourced work, the post-apartheid workplace still displays a racialized and gendered bias. Wits has been unsuccessful even in addressing the racial and gender imbalances in its cadre of academic staff. On 31 March 2002, only 22.4% of permanent academic staff at Wits was black and 43.1% was female. Furthermore, only 13.7% of senior permanent academic staff was black and 30.2% was female. Of the University's 118 professors, only 10 were black, of which only one was a black woman (University of the Witwatersrand 2002).

[27] Interview with Elizabeth, 5 April 2004.

[28] The Wits HR director was explicit that matters relating to the cleaners are not handled by his directorate, or that of Industrial Relations at Wits, but by the Finance Directorate (interview with Human Resources Director, 18 April 2004).

[29] Interview with Wits Human Resources Director, 18 April 2004.

[30] Significantly, this is the same group upon which the burden of social reproduction under apartheid generally fell, for it was older African women who were restricted to the homelands—often taking care of their grandchildren—whilst younger men were allowed to migrate into "white South Africa" to work in manufacturing or mining.

[31] Interviews with Joanne, 8 April 2004; Nozipho, 6 April 2004; Lindiwe, 21 April 2004; and Susan, 5 April 2004.

[32] Observation at Maria's home, May 2004.

[33] Statistics South Africa (2002:53) shows that African women and men spend the most time on unpaid housework compared with the other "racial" categories; 230 and 83 minutes a day, respectively. However, the disparity between the two genders of the African category is great; African women spend almost two and a half hours (147 minutes) per day more on unpaid household tasks. In effect, the African household requires 20–30% more time than colored, Indian and white households. This difference reflects the lack of resources available and used by African households (such as paid domestic work and child care), the quality of housing provided (the size of homes, running water and electricity), and the nature of the household itself (extended family networks as opposed to nuclear families).

[34] Interviews with Elizabeth, 5 April 2004; Joanne, 8 April 2004; Felicilty, 7 April 2004; Phumzile, 5 April 2004; Zanele, 7 April 2004; Nozipho, 6 April 2004; Brenda, 6 April 2004; Lindiwe, 21 April 2004; Susan, 5 April 2004.

[35] Observation at Maria's home, May 2004.

[36] Interview with Joanne, 8 April 2004.

[37] All the workers complained of the effects of the chemicals in their cleaning materials. None of the women were provided with masks or gloves. Dinah was on one-month's sick leave and her doctor advised her not to use the same chemicals when cleaning. Six of the women stated that the chemicals affected their skin and chests, and gave them cramps (see endnote 34). However, management did not respond to any of their requests to change the cleaning materials and refused to pay for the medical expenses related to these complaints or for the expenses of two of the women who hurt an arm and a finger whilst operating heavy floor polishers. One worker was told: "if I couldn't do my job, then I'd better go and get a doctor's note. Then I can stay at home forever. The government will then pay me for staying at home because Supercare will not" (interview with Phumzile, 5 April 2004).

[38] The South African public health system is overburdened because of the HIV-AIDS pandemic. The provision of medication for chronic illness has been scaled down and the cost for working people has increased threefold in the past five years. Since 2003, this means that chronic sufferers have to attend local clinics once a month at a charge of R35. Before these changes in policy, patients visited hospitals on a three-monthly basis at a charge of R13 per visit (interviews with Maria, 7 April 2004 and Joanne, 8 April 2004).

[39] Observation at Maria's home, May 2004.

[40] Various interviews. See endnote 34.

[41] Interviews with Maria, 7 April, 14 April 2004, 13 May and 14 May 2004; observation at Maria's home, May 2004.

References

Adler G, Bezuidenhout A, Buhlungu S, Kenny B, Omar R, Ruiters G and Van der Walt L (2000) "The Wits University support services review: A critique." Unpublished report

Adler G, Bezuidenhout A and Omar R (2000) Expert service, or a way to avoid unions? *Business Day* 7 June

Baskin J (1991) *Striking Back: A History of COSATU*. Johannesburg: Ravan Press

Bernstein D (1986) The subcontracting of cleaning work: A case in the casualization of labor. *Sociological Review* 34(2):396–422

Bertelsen E (1998) The real transformation: The marketization of higher education. *Social Dynamics* 24(2):130–158

Bezuidenhout A, Godfrey S, Theron J and Modisha G (2004) "Non-standard employment and its policy implications." Report submitted to the Department of Labour, South Africa

Burawoy M (1998) The extended case method. *Sociological Theory* 16:4–33

Cheadle H and Clarke M (2000) *National Studies on Workers' Protection: South Africa*. Geneva: International Labour Office

Clarke M, Godfrey S and Theron S (2003) *National Studies on Workers' Protection: South Africa*. Geneva: International Labour Office

Cock J (1980) *Maids and Madams: Domestic Workers Under Apartheid*. Johannesburg: Ravan Press

Crankshaw O (1997) *Race, Class and the Changing Division of Labour under Apartheid*. London: Routledge

Elson D (1999) Labor markets as gendered institutions: Equality, efficiency and empowerment issues. *World Development* 27(3):611–627

Hart G (2002) *Disabling Globalisation: Places of Power in Post-Apartheid South Africa*. Pietermaritzburg: University of Natal Press

Herod A (2001) *Labor Geographies: Workers and the Landscapes of Capitalism*. New York: Guilford Press

Humphries J and Rubery J (1984) The reconstitution of the supply side of the labour market: The relative autonomy of social reproduction. *Cambridge Journal of Economics* 8:331–346

Johnstone F A (1976) *Class, Race and Gold: A Study of Class Relations and Racial Discrimination in South Africa*. London: Routledge and Kegan Paul

Kenny B (2003) Labour market flexibility in the retail sector: Possibilities for resistance. In T Bramble and F Barchiesi (eds) *Rethinking the Labour Movement in the New South Africa* (pp 169–183). Aldershot: Ashgate

Kenny B and Bezuidenhout A (1999) Contracting, complexity and control: An overview of the changing nature of subcontracting in the South African mining industry. *Journal of the South African Institute of Mining and Metallurgy* 99(4):185–191

Kenny B and Webster E (1999) Eroding the core: Flexibility and the re-segmentation of the South African labour market. *Critical Sociology* 24(3):216–243

Molete M and Hurt K (1997) *Nehawu History—The Unfinished Story: The History of The National Education, Health, and Allied Workers' Union, 1987–1997*. Johannesburg: Nehawu

Moodie T D (1994) *Going for Gold: Men, Mines and Migration*. Johannesburg: Witwatersrand University Press

Mosoetsa S (2003) Re-emerging Communities in Post-Apartheid South Africa: Mpumalanga Township, Kwa-Zulu-Natal, Durban. Annual Workshop of the Crisis States Programme, 14–18 July, University of the Witwatersrand

Murray B K (1982) *Wits: The Early Years*. Johannesburg: Witwatersrand University Press

National Education, Health, and Allied Workers' Union (NEHAWU) (1985) "Perceptions of Wits (Worker's Perspective)." Unpublished report. Johannesburg: NEHAWU

Ngcobo N (2004) Contractors treat cleaning mothers brutally. *Wits Student* March:1, 3

Pearson R (2000) Rethinking gender matters in development. In T Allen and A Thomas (eds) *Poverty and Development into the 21st Century* (pp 383–402). Milton Keynes: Oxford University Press

Peck J (1996) *Work-Place: The Social Regulation of Labor Markets.* New York: Guilford Press

Pocock B (2003) *The Work/Life Collision.* Leichardt: Federation Press

Puech I (2004) Le temps du remue-ménage. Conditions d'emploi et de travail de femmes de chambre. *Sociologie du Travail* 46(2):150–167

Republic of South Africa (2004) Amendment of Sectoral Determination 1: Contract cleaning. *Government Gazette* 473(27029):1–4

Seidman G (1994) *Manufacturing Militance: Workers' Movements in Brazil and South Africa.* California: University of California Press

Shear M (1996) *Wits: A University in the Apartheid Era.* Johannesburg: Witwatersrand University Press

Silver B (2003) *Forces of Labor: Workers' Movements and Globalization since 1870.* Cambridge: Cambridge University Press

Statistics South Africa (2002) *Women and Men in South Africa: Five Years On.* Pretoria: StatsSA

Theron J (2003) Employment is not what it used to be. *Industrial Law Journal* 24:1247–1282

Theron J with Godfrey S, Lewis P and Pienaar M (2004) *Labour Broking and Temporary Employment Services: A Report on Trends and Policy Implications of the Rise in Triangular Employment Relationships.* Pretoria: Department of Labour

United Nations Development Programme (UNDP) (2004) *South Africa: Human Development Report 2003—The Challenge of Sustainable Development: Unlocking People's Creativity.* Geneva: UNDP

University of the Witwatersrand (2002) Second Employment Equity Progress Report (for the twelve-month period June 2001 to May 2002). Johannesburg: University of the Witwatersrand. http://www.wits.ac.za/tee/2002_report.htm (last accessed 21 October 2005)

Van der Walt L, Bolsmann C, Johnson B and Martin L (2003) Globalization, the market university and support service outsourcing in South Africa: Class struggle, convergence and difference, 1994–2001. *Society in Transition* 34(2):272–294

Von Holdt K (2003) *Transition from Below: Forging Trade Unionism and Workplace Change in South Africa.* Durban: University of Natal Press

Von Holdt K and Webster E (2005) Work restructuring and the crisis of social reproduction: A southern perspective. In E Webster and K Von Holdt (eds) *Beyond the Apartheid Workplace: Studies in Transition* (pp 3–40). Durban: University of Kwa-Zulu-Natal Press

Wallerstein I, Martin W G and Vieira S (1992) *How Fast the Wind? Southern Africa, 1975–2000.* Trenton, NJ: Africa World

Webster E (1985) *Cast in a Racial Mould: Labour Process and Trade Unionism in the Foundries.* Johannesburg: Ravan Press

Webster E and Omar R (2003) Work restructuring in post-apartheid South Africa. *Work and Occupations* 30(2):194–213

Wolpe H (1972) Capitalism and cheap labour-power in South Africa: From segregation to apartheid. *Economy and Society* 1(4):425–456

Zlolniski C (2003) Labor control and resistance of Mexican immigrant janitors in Silicon Valley. *Human Organization* 62(1):39–49

Chapter 4

Restructuring the Architecture of State Regulation in the Australian and Aotearoa/New Zealand Cleaning Industries and the Growth of Precarious Employment

Shaun Ryan and Andrew Herod

Introduction

Whereas until fairly recently the economies of Australia and Aotearoa/ New Zealand were dominated by agriculture and manufacturing, today some 70% of the working population in both countries is employed in services (OECD 2002). The majority of these new service jobs are low-paying, low-tech jobs, such as professional cleaning (Sassen 2001; Thompson, Warhurst and Callaghan 2001). This growth in service work has been accompanied by dramatic changes in the world of work organization, in both the service and non-service sector. Thus, employers have increasingly encouraged part-time and casual work whilst government has encouraged labour market "flexibility" and economic individualism and has sought to erode national systems of industrial regulation and to undermine institutions of collectivism (such as unions). In this regard, Australia and New Zealand have followed trends evident in other industrialized nations, including the USA and the UK.

Significantly, Australia and New Zealand share many labour market similarities—both have relatively similar proportions of female labour market participation and part-time employment, and their wage structures and unemployment rates are also comparable. Such similarities result from a shared cultural and political heritage and having experienced similar paths towards economic development (Brosnan and Walsh 1998:23). They also share a certain uniqueness in that they are the first two countries in the world to have adopted a national system of compulsory conciliation and arbitration (Barry and Wailes 2004:431), a system that makes it easier for weak unions to protect workers because they can use arbitration courts to force employers to improve pay and working conditions, rather than having to rely upon lengthy industrial disputes.[1] However, whereas such systems were at the core of both countries' industrial relations systems for most of the 20th century, in the late 1980s the New Zealand Labour government adopted an aggressive policy of neoliberal transformation which involved dismantling the system of arbitration which had existed since the 1890s. This has been followed more recently by the Howard administration's efforts in Australia to reimagine labour market regulation along neoliberal terms.

In this paper, then, we explore how the dismantling of the architecture of labour market regulation is shaping the contemporary development of the cleaning industry. Drawing upon interviews with cleaners, managers and industry association officials, we examine particularly how employment precariousness is being introduced into the lives of cleaners through a variety of neoliberal practices, including those of outsourcing, work intensification, and the privatization of government services—all of which have been facilitated by a recasting of the state's regulatory relationship to labour markets through the dismantling of systems of compulsory arbitration. First, we provide an overview of the arbitration system whereby wages and conditions of employment were fixed nationally. We then highlight something of the industry's historic organization as a means by which to contrast contemporary developments—for instance, whereas the system of compulsory arbitration encouraged the proliferation of fragmented and sometimes overlapping unions and allowed a small number of large companies to dominate, now larger numbers of small companies are emerging and unions are being forced to explore new ways of operating (including merging) so as to protect workers. We follow this with an examination of how the changing regulatory environment is encouraging the growth of precarious employment in cleaning and a decentralization of industrial relations as greater emphasis is placed upon enterprise-level collective bargaining to set employment remuneration. Finally, we explore some of the implications such changes have for the possibilities for cleaners' unionization and collective bargaining.

Establishing and Dismantling Systems of Compulsory Arbitration in Aotearoa/New Zealand and Australia

The latter part of the 19th century in both New Zealand and Australia was marked by bitter conflicts between capital and labour which threatened both nations' economic futures. In response, governments moved to implement systems in which arbitration courts would rule on the particulars of any difference of opinion between employers and their workers and would establish minimum wage rates and other conditions of employment (called "Awards"). In 1894 New Zealand became the first to adopt such a scheme when the government enacted the Industrial Conciliation and Arbitration Act requiring trade unions and employers—either individually or as part of a union federation or employers association—to register as the appropriate parties for collective bargaining purposes and to engage in arbitration proceedings if they failed to reach agreement. By the mid-1960s, some 42% of all wage and salary earners were subject to awards (McLintock 1966).

Likewise, in Australia several particularly disruptive strikes in key export sectors during the 1890s led the federal parliament in 1904 to pass the Conciliation and Arbitration Act establishing an Arbitration Commission as an umpire between employers and unions. This Commission was supposed to bring about, in the words of Prime Minister Alfred Deakin, a "People's Peace" by removing the need for industrial action through ensuring workers were paid a fair wage. The Act quickly led to a system of national wage regulation and industrial entitlements and, although amended several times, its basic provisions remained in place until the 1980s.

Recently, however, pressures for change have mounted. In New Zealand, a dramatic spike in the number of industrial disputes (from an average of around 70 per year in the mid-1950s to more than 550 per year by the mid-1970s; Statistics New Zealand 2000) and a desire to encourage the amalgamation of the myriad small unions which had flourished under the arbitration system into a smaller number of larger, better funded unions led the Labour government in 1984 to pass the Industrial Relations Amendment Act, thereby making arbitration voluntary in the private sector and encouraging settlement of disputes through collective bargaining. This was followed in 1987 by requiring unions to have a minimum of 1000 members to register as a recognized entity. The process of "reform" was completed in 1991 when the new National government passed the Employment Contracts Act, bringing to an end the practice of regulating labour markets through compulsory arbitration and the Arbitration Court's issuance of national awards. In its place, in an attempt to bring about "an efficient labour market" (Statistics New Zealand 2005), the government instigated a system in which employees and employers could choose either to negotiate an individual employment contract or to be bound by a collective employment contract. The 1991 Act, then, resulted in "the removal of all remaining vestiges of the arbitration system in New Zealand" (Wailes and Ramia 2002:85).

In Australia, by way of contrast, the arbitration system remained largely in place throughout the 1980s, with reform—also in response to a spike in strikes—involving "modernizing" existing institutions, rather than replacing them wholesale. This resulted in the repeal of the 1904 Act and passage of the Industrial Relations Act of 1988, which replaced the Arbitration Court with the Australian Industrial Relations Commission (AIRC) which, for the most part, was similar in nature to the old Arbitration Court, though it sought increasingly to link awards to productivity improvements (AIRC 2005; Wailes, Ramia and Lansbury 2003:624–625). Such reform was undertaken under the aegis of a "Prices and Income Accord" between the ruling Australian Labor Party and the Australian Council of Trade Unions (ACTU). However, the federal election of the Liberal/National Party coalition government in 1996 and the coming to power of Prime Minister John Howard resulted in the Accord's collapse and the Howardites soon began to institute legislation to dismantle the award system.

One of the most important pieces of legislation the Howard government passed was the 1996 Workplace Relations Act limiting the practice whereby the AIRC would establish awards granting all workers in an industry the same minimum conditions of employment and wages whilst leaving them free to negotiate through collective bargaining wages and conditions better than the award. Not only did the 1996 Act reduce awards' scope but it also allowed for individualized employment contracts—known as Australian Workplace Agreements (AWAs)—between employers and employees which, like collective bargaining agreements, can (with some exceptions) override existing award conditions. Significantly, though, whereas enterprise agreements could only result in wages and conditions which were better than any particular award, AWAs can reduce wages and conditions below an award level—whereas AWAs must leave workers no worse off "as a whole" relative to an industrial award, in practice some workers have ended up being paid as much as 25% less than the appropriate award (ACIRRT 2002; Fairgo.nsw.gov.au 2005).

This realignment between the two systems, though, may be short lived, as the Labour-led coalitions which have governed New Zealand since 1999 have sought to extend multi-employer bargaining and employment protection to contract workers, especially cleaners. In Australia, on the other hand, the Howard government has moved to introduce even greater "flexibility" into labour markets by passing the 2005 Workplace Relations Amendment (Work Choices) Bill, which will further deregulate labour markets. Put another way, whereas in New Zealand the national government appears to be pulling back on some of the neoliberal agenda which has dominated legislation for the past twenty years, in Australia the federal and many state governments are proceeding with even greater neoliberalization of labour markets. This is significant, given that what happens in the Australian economy has a much greater impact on New Zealand than vice versa.

The result of all this is that, despite recent developments in New Zealand, the erosion of national wage and conditions minima afforded by awards means that employment conditions in Australasia are rapidly becoming less secure for many workers. In this regard, the concept of "precarious employment" (cf Tomic, Trumper and Dattwyler, this collection) provides a useful analytical tool for exploring labour market changes. Such precariousness, Burgess and Campbell (1998) have argued, can result from two main processes— the deliberate erosion of rights and protections, and the expansion of contingent (ie part-time and temporary) employment. Ominously, the Australasian cleaning industry is enduring both, with cleaners seeing the emergence of a "regime of [employment] insecurity, [one] featuring individualisation [and] the organisational fragmentation of the workplace, [changes which are increasingly] coupled with flexibility in hours worked and the length of employment" (Allen and Henry 1996:67).

Characteristics of the Australasian Industry

In order to comprehend contemporary changes being wrought in the Australasian cleaning industry it is first necessary to outline how the industry operated throughout much of the 20th century. This is important not just for understanding the impact domestically of the restructuring of the New Zealand and Australian systems of labour market regulation but also for appreciating the role played by the Australasian industry in transmitting globally employment and organizational practices, for although the Australasian industry is small in international terms, claiming around 5% of the global industry, it has made large contributions in other ways—for instance, many of the industry's pioneering innovations originated in Australasia, including backpack vacuum cleaners and rotary suction polishers (Jeffs 2001). Such innovations in techniques and equipment are partly due to the intensely competitive nature particularly of the Australian industry but also contribute to the extraordinary rate of productivity claimed by it—Australian firms are "acknowledged world leader[s] in productivity in CBD high rise" buildings and many claim they can clean 1000 m^2 per hour (BOMA 1990; Jeffs 2001), compared to only 300–400 m^2 per hour in North America, although such levels of productivity are generally achieved at the expense of quality and the erosion of cleaners' conditions of work (Ryan 2001).[2]

In recent years, many firms have become highly diversified and provide a range of services under the practice of "facility management", including housekeeping (cleaning, food, security, laundry) and maintenance functions (grounds keeping, window cleaning, plumbing). Notably, such practices were pioneered in New Zealand in the late 1950s by the firm Crothalls, though they have subsequently become virtually standard practice globally as companies overseas have copied the Crothalls model and as Australasian firms (which were some of the first globally to expand beyond their home nations) have bought local firms overseas. The two biggest Australian firms are large by international standards—Tempo Services claims to hold the world's largest cleaning contract, employing over 3500 staff to clean 1970 public sector sites in New South Wales (Tempo 1999), whereas Berkeley Challenge (formerly part of the Crothalls empire but now part of Spotless Services) employs 9000 cleaners across 2400 different contracts (Spotless 2002). The industry is also highly fragmented, partly as a result of the fact that the government in both Australia and New Zealand has had a key role in its development by virtue of the large number of cleaners employed by disparate public sector departments and institutions, particularly hospital and education departments (Ryan 2001)—at its peak, for example, the New South Wales Government Cleaning Service employed around 12,000 cleaners (Jensen and Liebenberg 1995).

The Australasian industry is highly competitive, with average profit rates in the range of 1–5% per annum (Brosnan and Wilkinson 1989:46; Jeffs 2001). Indeed, such is the nature of this competition that firms often bid on contracts at a loss, hoping to recover costs by subcontracting work to cheaper providers, by cutting cleaners' hours and reorganizing work, or by creating "specials" whereby the firm encourages the client to purchase additional cleaning above and beyond contracted specifications. The individualization of employment relations in New Zealand and in some Australian states in the early 1990s significantly exacerbated this competition, resulting in a situation in which, according to the industry journal *Inclean* (Oct–Nov 2004:15), "numerous quoted tender prices, particularly in NSW, [were] way below accepted industry benchmarks", although one of the largest firms has recently announced that it will abandon price-driven tendering in favour of delivering a cleaning package based upon quality and service (*Inclean* Dec 2004:35)—a decision that probably has much to do with the fact that larger firms have higher overheads and face more public scrutiny and so have experienced considerable difficulties in competing on price with smaller and/or less scrupulous firms.

Structurally, the industry is divided between a small number of large firms and a proliferating number of small "mum and dad" enterprises. In Australia 48% of cleaning enterprises are sole proprietorships or partnerships, whilst in New Zealand around 68% of cleaning enterprises are considered "self-employed" businesses. Although the industry is dominated numerically by small firms (89% of New Zealand cleaning enterprises and 80% of Australian ones employ fewer than ten people), until recently it has been the larger cleaning firms that have set the industry's tone and pace—in Australia in 2000, for instance, enterprises employing 100 or more staff accounted for fewer than 2% of all cleaning firms but employed some 55% of the workforce and generated 52% of industry income (ABS 2000), with figures for New Zealand being similar in terms of income, though there large firms have a smaller share of total employment. Historically these large firms have tended to control industry groups and to set the pattern in terms of wages and working conditions. Furthermore, as a result of their size—two of the largest cleaning firms (Spotless and Tempo) are amongst the largest 20 firms in all of Australasia—and industry dominance they have traditionally been able to secure significant numbers of public sector cleaning contracts.

The relationship between small and large firms has traditionally developed along parallel lines, with some smaller firms entering into symbiotic relationships with larger firms. In New South Wales, for instance, some of these smaller firms have grown through serving as subcontractors to larger cleaning firms, although changes to state awards have attempted to regulate this practice by holding the principal contractor liable for the actions of the subcontractor (including concerning pay and conditions of employment), with the result that, historically, principal contracting firms have not been able legally to use subcontracting firms that offer wages and terms of employment less than those provided in the award. This situation has been reinforced by the fact that Australasian unions have a long history of prosecuting subcontractors who do not live up to the law.

As labour markets began to be deregulated in the 1990s there was a marked increase in the number of firms in the industry—in Australia the number increased from 4181 in 1987–88 to 5938 in 1998–99 (ABS 1990, 2000), whilst in New Zealand there were 2349 cleaning enterprises in 2004, a substantial increase from the 1887 recorded in 2000 (Alexander 2005). Two factors appear to have accounted for this rise. First, larger firms' higher operating costs and overheads have encouraged some to reconsider their relationship with competing smaller firms and many have begun to franchise operations, necessitating the hiring of cleaners and the creation of additional small firms to take advantage of such opportunities (Meagher 2003). Thus, in 2004, Tempo Services, one of the world's largest cleaning firms, began franchising and offering partnerships to smaller operators in a bid to increase its 15% share of the Australian market. Likewise, Jani-King, a US firm, has established 630 franchises across Australasia (*Inclean* Oct–Nov 2004:14–15). In New Zealand particularly, the development of franchising has led to the emergence of large numbers of small enterprises—such as husband and wife teams working small contracts—who have been able to carve out a niche cleaning schools, small offices and shops (Verkerk 2005). This has been fuelled by the fact that New Zealand immigration laws require migrants entering under the business investment programme to invest at least one quarter of their funds into a local enterprise, whilst the fact that they do not require a large amount of investment to set up has meant that cleaning franchises have been popular amongst Asian migrants, particularly those from China, as a way of starting their own business (Verkerk 2005).

Second, growth has arisen from outsourcing in the private sector and compulsory tendering and contracting in the public sector. This is especially evident in Australia. Thus, in New South Wales over 12,000 jobs were transferred from the Government Cleaning Service to the private sector between 1989 and 1994, of which some 7800 were transferred directly to private cleaning contractors. The extent to which cleaning has been outsourced to private firms is evident in the Australian Workplace Industrial Relations Survey, which reported that one-third of firms contracted services, including cleaning, between 1990 and 1995 (ACIRRT 1999:142), with current indications being that this trend is continuing.

The result of these transformations in the industry's structure has been a significant increase in the size of the workforce over the past decade, with cleaning now being the third largest employer in Australia (behind sales assistants and secretaries and personal assistants) (Culley 2002). Hence, between 95,000 and 130,000 people were employed on contracts by cleaning firms in the early 2000s, whilst census figures put the total number of cleaners at 188,000 (Culley 2002).[3] Figures for New Zealand (with a population one-fifth the size of Australia's) show a total of 12,195 persons working in the industry (Verkerk 2004). Women make up around 60% of the New Zealand cleaning labour force and around half of the Australian one.

Changing Regulatory Environments and the Growth of Precarious Work

Despite the variations in the structure of the industry's firms, the lack of any system of centralized recruitment strategies and the almost complete absence of formalized human resource management techniques, there is, nevertheless, a certain degree of similarity in the ways in which the industry as a whole is responding to the changing regulatory environment in which it now finds itself (Allen and Henry 1996). Part of the reason for this is that industry culture and labour practices are reinforced by a considerable circulation of managers amongst larger cleaning firms. However, the convergence of work practices and the organization of work in the industry also needs to be read in light of the increasingly competitive nature of the wider industry. Specifically, because labour is the most significant cost faced by cleaning firms, the dismantling of the award system and the growth of individualized contracts have meant that labour cost-cutting practices adopted by a number of firms are spreading widely and quickly, leading to a downward spiral of wages and conditions (LHMU 2003:14). Given that New Zealand abolished its award system first, cost-cutting strategies are already fairly advanced there, although they are deepening and spreading in Australia too. Thus, Western Australia's and Victoria's experiments with labour market deregulation in the 1990s encouraged individual contracts allowing small contractors to undercut standard wages and conditions—often by coercing cleaners into signing "voluntary" individual agreements—with the result that·many have experienced significant cuts in wages, as much as A\$40–50 a week (Creed 2005). This is noteworthy because it is speculated by many critics that the federal Work Choices legislation—which unions claim was inspired by the Western Australian experience, which itself drew inspiration from New Zealand's Employment Contracts Act—will have a similar effect nationally.

In an industry dominated by "champagne cleaning specifications" at "flat beer prices" (BOMA 1990), the main consequence of the changed regulatory environment has been an increase in employment risk and insecurity for cleaners (Allen and Henry 1996; 1997) as cleaning firms have sought to maintain profitability at the expense of their workforce. This is particularly so because government and private companies tend to view cleaning as a necessary expense to be minimized—rather than as an activity integral to facilitating the company's main pursuit—and have increasingly sought cost reductions from cleaning firms (LHMU 1999). As a result, cleaners rank amongst the lowest paid workers in Australasia (Brosnan and Wilkinson 1989; Buchanan and Watson 1997; Watson et al 2003), often relying upon statutory increases based upon Minimum Wages Legislation in New Zealand and Living Wage cases undertaken by the ACTU in Australia to maintain their incomes (Buchanan and Watson 1997).

Competitive pressures and the growing reliance upon enterprise-level collective bargaining rather than the centralized awards system to determine wages and conditions (Buchanan and Callus 1993) are resulting in three significant changes in the lives of cleaners. First, the economic landscape is being made much more variegated in terms of wage rates and workplace protections. Whereas previously a uniform set of wages or regulations might be expected to exist nationally or across a state, now there is the possibility for a much greater degree of spatial variability as wages and working practices are determined at the level of individual companies or local workplaces. Whereas the award system's provision of a wage floor encouraged firms to compete largely on the basis of more efficient work organization (which might allow cost savings), the deliberate atomization of the economic landscape has encouraged direct wage competition between firms.

Second, the search for flexibility has meant that many previously full-time jobs are being converted to part-time or casual work, resulting in short, insecure hours, intensified work and loss of pay and entitlements for workers (Allen and Henry 1996; Bernstein 1986; Brosnan and Wilkinson 1989; Leonard 1992; Watson et al 2003). Thus, it is estimated that 60–70% of the Australasian cleaning workforce has at some point been engaged on a part-time basis. Furthermore, whereas the Building Service Contractors Association of Australia (Jeffs 2001) contends that average hours of work are around 15 hours per week, the Australian Liquor, Hospitality and Miscellaneous Workers' Union (LHMU 1999) claims that one-third of the Australian cleaning labour force works fewer than 15 hours per week and half of all cleaners work fewer than 30. The 2001 New Zealand census reveals that around half of all cleaners work fewer than 30 hours per week, with half of these averaging 14 hours or fewer (Verkerk 2004). Moreover, across Australasia even regular part-time cleaning work is becoming "irregularized" as more cleaners are employed on a part-time basis with no security of hours (LHMU 2003:8). As a result, in both countries multiple job-holding is becoming increasingly common as cleaners try to secure additional hours (LHMU 2003; Robinson 2005). Meanwhile, the upwards recalculation of workloads that typically takes place when contracts are transferred to new parties (Ryan 2001) has implications for occupational health and safety, and the cleaning industry has witnessed growing numbers of injuries in recent years (Alcorso 2002; also Søgaard et al and Seifert and Messing, this collection).

Third, through contracting out for cleaning services businesses are able to divest themselves of economic risks. Thus, whereas previously employers might worry about having to continue paying employees who were "on the clock" when there was too little work to keep them busy, now it is employees who must worry about what they will do when there is too little work. By moving to a more contingent workforce, then, firms are able to maintain their operations without having to commit long-term to employees and workers' job security effectively becomes limited to the tenure of the contract between the business and the cleaning firm. This has profound consequences for most cleaners, because, although some contracts are of up to three years' duration (usually for government or prestige contracts in some areas), contracts with small- and medium-sized firms are typically month-by-month or subject to 30 days' notice (interview with owner of cleaning firm 12 February 2003).

Such changes clearly add to cleaners' labour market precariousness, especially because any more widespread use of AWAs is likely to encourage price wars that will undoubtedly encourage the larger companies which presently bargain collectively to offer individual contracts as a way to reduce wages and retain contracts (Robinson 2005). Furthermore, it is likely that those workers who have few other options—what Coyle (1985:7) has called the "trapped" sections of the workforce—and so who are forced into cleaning by necessity rather than enter it by choice will be impacted most. Such workers, many of whom the industry itself has actively sought out as it has striven to make employment more contingent, include women who have domestic responsibilities which preclude other types of employment, ethnic minorities looking for jobs that will give them entry into the formal labour market (Aguiar 2001; Meagher 2003), non-native-English speakers, those with little formal education or those from overseas whose educational qualifications are not recognized in Australia and New Zealand, and the old (often, workers who have been made redundant as a result of deindustrialization) (see Alcorso 1991, 2002; Fraser 1997; Gold 1964; Leonard 1992).[4] This has been exacerbated by the entry into the market of immigrant-based cleaning firms who take advantage of immigrant networks to find workers but who then use immigrants' need for work, their frequent lack of knowledge of their legal rights and their unwillingness often to speak out (whether out of fear of the authorities or out of a sense of ethnic loyalty to their employers) to exploit them (Robinson 2005). Thus, as one senior manager explained in relation to the New South Wales cleaning industry:

> From my experience cleaning companies tend towards employing one particular race ... [In] my old company we employed ... many Thai people. We had 600 at one stage. And why? Because we had Thai-speaking supervisors, so the more people from one race the easier was the communication between the staff, as well as from supervisors on down. One of the major companies was built on Greek people, and so it goes on (interview with John, 12 February 2003).

Unionism in an Age of Precarious Employment

Unions representing cleaners in New Zealand and Australia—the Service and Food Workers' Union of Aotearoa (SFWU) and the LHMU—are amongst the largest unions in both countries and historically have enjoyed some considerable political clout, with former officials from these unions having held important positions in recent years on the New Zealand Labour-led coalition government and the state Labor governments in Australia. Despite this, across Australasia union representation amongst cleaners is, at best, patchy, with rates of unionization in cleaning tending to be significantly lower than the national average—whilst total union density is 21.4% in New Zealand (May, Walsh and Otto 2004), across the cleaning industry it is only around 12% (*Inclean* Apr–May, 2004:10), whereas in Australia, though it is not known precisely what is the density in the industry, most commentators assume it is significantly lower than the 23.1% national rate of union membership (Cooper 2004). In both countries, membership is strongest in larger cleaning firms and worksites and within the public sector. Thus in 2002 one of the largest Australian cleaning firms had a density of 27% (interview with HR Manager, 17 April 2002), whereas union membership among New South Wales school cleaners is around two-thirds of cleaners (interview with NSW LHMU officials, 12 July 2000). By way of contrast, the vast majority of the "mum and dad" firms now emerging are non-union operations.

The growth in the number of "mum and dad" firms and in the outsourcing of government cleaning contracts has significant consequences for unions. Thus, whilst they have had some success in limiting outsourcing—in New South Wales, for example, a protracted campaign drawing inspiration from the Service Employees' International Union's Justice for Janitor's campaign in the United States has been successful in protecting cleaners' hours and jobs by preventing the subcontracting of 7000 jobs from NSW schools (*LHMU Union News*, 11 November 2005)—without reinvigorated organizing campaigns the level of industry unionization will probably fall in future years as governments subcontract out more work and as the larger firms face growing competition from smaller firms. This is particularly the case because although the subcontracting of government work to private companies could potentially transfer cultures of militancy from the government to the private sector as formerly government-employed workers take their traditions of trades unionism with them into their new contracts (cf Wills 1998), in actuality the fact that many of the new and small cleaning contractors employ workers (often immigrants) with little history of involvement in the industry or with unionism means that much of the industry is being reinvented largely as a non-union operation.

For their part, the unions have historically concentrated their organizing activities upon larger employers, the result both of hoping to gain a greater payoff for their efforts and because only the large firms tend to have specialist human resource (HR) managers, with whom many union leaders feel more comfortable negotiating. This strategy, though, appears to be problematic, both because the fastest rates of employment growth are occurring in smaller firms and because experience suggests it is smaller firms which are most likely to circumvent collective agreements (though large firms are by no means guiltless!). Furthermore, because only the very largest firms tend to have specialist HR staff, somewhat arbitrary employment relations have developed within the small-firm sector as regimes of labour control and agreements between firms and their employees have often been determined on an *ad hoc* basis (Rees and Fielder 1992). These structural characteristics, combined with the fragmented and dispersed nature of cleaning, the irregular hours of work and the poor English skills amongst many cleaners, have made unions' policing of wages and conditions difficult (interview with LHMU organizer, 4 April 2003), particularly in smaller firms where cash-in-hand payments are common.

The growth in "mum and dad" firms facilitated by the ending of the award system in New Zealand, and its severe curtailing in Australia, then, have dramatically transformed the arena in which unions operate in both countries. Whereas for most of the 20th century pay and conditions of employment in both Australia and New Zealand were regulated through awards operating at the national or state level, collective bargaining and individualized bargaining have increasingly determined workers' pay and conditions of employment. This has had momentous implications. For instance, after the system of arbitration in New Zealand was ended by the Employment Contracts Act (Oxenbridge 1999), collective bargaining and union membership fell dramatically and individual agreements became the primary determinant of pay and conditions across the economy for workers in small, private sector service industries (Hamberger 1995:8), although the SFWU subsequently managed to negotiate a multi-employer collective contract with the larger firms, one which then became a benchmark for non-union firms (SFWU *News*, 7 December 2001).

Given that the contents of individual agreements made between employers and employees are not publicly available, it is not readily apparent how cleaners who fell outside the scope of this collective cleaning agreement have fared. However, reports from the SFWU do indicate that "mum and dad" enterprises and smaller and provincial cleaning firms have been underpaying workers, with the result that, despite the existence of a multi-employer agreement prescribing NZ$10.60 per hour (a rate only marginally above the NZ$10.25 minimum adult wage), wages in the New Zealand cleaning industry have stagnated—in 2001 nearly 14,000 of the 32,700 people employed as cleaners earned less than NZ$10,000 per year and almost 25,000 (some 75%) earned less than NZ $20,000 (KiwiCareers 2004; see also Kalb and Scutella 2003).

Nevertheless, despite the problems cleaners and their unions face, some progress has been made on their behalf in New Zealand. Thus the 1999 return to power of the Labour-led coalition government and its passing of the Employment Relations Act (2000) did increase legislative protections for cleaners. Equally, an amendment to the Act

in 2004—one strenuously opposed by employers—specifically identified cleaners and caretakers as workers who would have the right to
transfer to a new employer on their existing terms and conditions of
employment if the business were sold, contracted out or if a contract
were lost to another contractor (see Tomic, Trumper and Hidalgo
Dattwyler in this collection for an examination of the opposite trend
in Chile). Additionally, if cleaners choose not to transfer to the new
employer, under certain conditions they are entitled to redundancy
payments (Employment Relations Service 2004:5). The implications
of this amendment are far-reaching and mean that employers may no
longer legally cut jobs and wages, nor erode hours or other conditions
of employment, upon the transfer of contracts, previously a common
cause of grievance. Nevertheless, the Act is deeply ambivalent.
Specifically, by allowing only registered unions to negotiate collective
agreements some cleaners who are unable to join or form unions
for whatever reason but who had previously negotiated with their
employers various kinds of loose consultative mechanisms may end
up being worse off. Furthermore, whereas the Employment Relations
Act reinstated the right of workers to engage in industrial action to
force firms to participate in multi-employer bargaining (a right taken
away in 1991), the Act has tended to encourage enterprise rather than
occupational unionism—that is to say, the focus has been on promoting unionism at the decentralized scale of the individual workplace
rather than at the more centralized scale of the industrial sector as a
whole (Barry and Walsh 2002). This makes it more difficult to engage
in the kinds of bargaining across whole labour markets—or segments
thereof—that has characterized the activities of, for instance, the
Justice for Janitors campaign in Los Angeles (see Savage 1998).

Given its federal system, the regulation and nature of employment
in Australia is more fragmented than it is in New Zealand. Whereas
in most Australian states cleaners are still employed under various
state awards which cover pay and conditions, other cleaners come
under the aegis of federal or enterprise agreements, whilst some are
employed on individual contracts. However, despite the fact that the
1996 Workplace Relations Act allowed individualized contracts in
the form of AWAs, collective agreements remain the norm across
most Australian states. Moreover, and perhaps somewhat surprisingly,
large cleaning firms have actually supported collective bargaining
and the maintenance of a common rule agreement regulating wages
and working conditions because, with 80–85% of contract costs being
directly related to labour, such firms have worried that the move to
AWAs will allow smaller firms with lower operating costs to undercut
them. Whilst the variety of employment arrangements across Australia
makes it difficult to generalize rates of pay with any accuracy,
rates for part-time cleaners (the majority of cleaners) in state awards

suggest significant regional variation across labour markets, from A$13.15 per hour in Western Australia to an average of A$15–16 in Victoria, Queensland, New South Wales and Tasmania. Furthermore, whereas there is no legal requirement in any state for continuance of employment upon a transfer in contract as there now is in New Zealand, the LHMU has managed to negotiate a series of "status quo" agreements across all states to provide some security, though because it operates as common agreement rather than law it is widely abused (interview with LHMU official, 4 April 2003).

The recent "reforms" of industrial relations legislation portend, however, momentous changes for workers and unions. Hence, whereas the Howardites have presented the 2005 Work Choices Act in the neutral language of giving workers "choice" and of "simplifying" industrial agreements by seeking to abolish the various state systems of industrial relations and replacing them with a national system of industrial relations, if one looks beyond such language, opponents argue, workers will find "a nasty right-wing hand-me-down ideology to Americanise our workplaces" (Beazley 2005) as the Act transfers the power to set minimum wages from the Industrial Relations Commission—made up of judges who have some independence of tenure and long seen as a protector of union rights—to a Fair Pay Commission made up of government appointees who will seek to "ensure that wage rises are underpinned by productivity improve-ments" (Commonwealth of Australia 2005) rather than reflect in-creases in the cost of living. Perhaps more significantly, the Act also abolishes unfair dismissal restrictions on firms employing up to 100 people, with the result that almost 70% of the Australian workforce will be excluded from unfair dismissal laws, whilst reducing the number of pay and conditions standards protected under the award system from twenty to just five (covering a minimum hourly wage rate, sick, parental and annual leave, and maximum weekly working hours). Furthermore, there is concern that many larger companies will now begin to break up their operations into multiple employment units of fewer than 100 employees so as to escape the unfair dismissal provisions. By enshrining such provisions in federal law, critics suggest, the Howard government hopes to limit the power of state legislatures to bypass the federal government by maintaining their own sets of labour market regulations.

Unions, then, have become increasingly concerned that the Work Choices Act is simply a way to encourage the adoption of AWAs so as to undermine employment and collective agreements and to reduce the minimum wage, thereby encouraging "flexibility" in the labour market, especially at its lower rungs. Specifically, they worry that the new legislation allows employers to change employees' work hours without reasonable notice, removes the "no disadvantage test" from collective and individual agreements, abolishes annual wage increases historically made by the Industrial Relations Commission (with the ultimate goal of reducing the minimum wage in real terms), delays wage increases under the National Wage Case for 1.7 million low-paid workers for a period of 18 months or more and allows the government to fine union representatives up to A$33,000 for including health and safety, training and other such clauses in collective bargaining agreements. Furthermore, by allowing the Minister for Employment and Workplace Relations to strike what s/he considers to be "prohibited content" from a collective bargaining agreement, they fear the Act will give the government control over every wage agreement negotiated in Australia. For opponents, though, perhaps the most insidious part of the Act is Section 104(6), which maintains that it is not considered "duress" for an employer to require new and existing employees to sign an AWA and that, despite government reassurances to the contrary, "it [will not be] difficult for employers to develop ways and means of applying pressure on current employees who are reluctant to sign an AWA, without being in technical breach of the legislation" (Australian Senate 2005).

Conclusion

Despite the move towards the individualization of contracts and the decentralization of industrial relations from national awards systems to bargaining at the firm or plant level in both countries, there are nevertheless several contradictions that are playing out in the restructuring of industrial relations systems in Australia and New Zealand. For example, whilst labour market deregulation and "flexibility" are generally seen as things which capital favours, in Australia many of the big cleaning firms have actually supported industrywide collective bargaining as a way to limit the ability of smaller firms to undercut them. Likewise, even as John Howard has suggested that the centralization of the old industrial relations system was "itself . . . the problem" (quoted in Wailes 1997:39) and that this must be corrected through the decentralization of collective bargaining to the level of individual workplaces, this process has itself required a centralization of control at the federal level that will allow the national government to limit the ability of the states to implement legislation which might prevent his plans for increasing labour market flexibility. Equally, despite having very different institutional contexts within which labour market deregulation has played out—in the case of New Zealand, no written constitution and a parliamentary system with a single legislative chamber, in the case of Australia, a federal system with a written constitution laying out the powers of the federal government—for much of the 20th century the two nations have followed similar industrial relations paths. At the same time, though, these different systems have shaped the introduction of labour market deregulation in quite particular ways—whereas the New Zealand system allowed an exceedingly rapid deregulation in the early 1990s, in Australia the Howard government has been more constrained in its attempts to pursue neoliberal deregulation and it has taken longer to introduce individual contracting in industrial relations (Wailes 1997).

The key questions, though, relate to what the changed architecture of labour market regulation means for the abilities of unions to organize and bargain collectively. Thus, in both Australia and New Zealand unions representing cleaners have argued that growing employment precariousness is the result of the dismantling of the old regulatory framework, a structural transformation which makes it difficult for them to make gains through collective bargaining. Furthermore, changes in the industrial relations framework in both countries—particularly the encouragement of decentralized bargaining—mean that contracts are increasingly being negotiated for individual workplaces rather than for the broader occupation/ industrial sector (LHMU 2003:11). This is exacerbated by the fact that the fractured nature of the cleaning industry and the proliferation of small firms tends to limit the effectiveness of collective bargaining to large firms, whilst the prevalence of individualized contracts and the lack of an effective enforcement mechanism amongst small "mum and dad" firms make it difficult to use such contracts as patterns across the industry, as has been the case at other times and in other industries and nations. Hence, although some 10,000 New Zealand cleaners are covered by the multi-employer collective agreement (SWFU *Our Voice* Autumn 2004) and many Australian cleaners work under state awards, what collective bargaining does exist is increasingly being undermined by the ability of non-signatories to offer contracts at lower costs to firms—an ability facilitated by the deregulation of labour markets and the abandonment of centralized wage fixing. The result is that cleaners in Australasia face low wages, short working hours and significant job insecurity, such that they are among the lowest paid workers in both countries.

Despite a worsening of the structural conditions within which they must work, all is not lost for unions representing cleaners. Thus, in a bold tripartite international campaign to reorganize cleaners in the commercial office sector, the SFWU in New Zealand and the LHMU in Australia have recently joined forces with the US Service Employees' International Union to pressure simultaneously employers in all three countries. Stemming from the SEIU's "Global Strength" commitment made at its 2004 national convention, the goal of such a partnership is to organize "whole markets in every part of a global employer's operations" (Crosby 2005). Significantly, unlike more traditional international solidarity efforts in which unions in one country give moral and sometimes financial support to those in other nations, the SFWU-LMHU-SEIU campaign involves the SEIU actually sending organizers and research staff to Australasia to work directly on organizing campaigns—including some who, in November 2005, helped the SEIU win what has been billed as the biggest organizing campaign in the southern United States in recent years when 5300 cleaners who work for the five largest cleaning companies in Houston, Texas, unionized. The hiring of more than 30 organizers and five researchers makes the SFWU-LMHU-SEIU campaign "probably the biggest growth campaign run in either country in over a hundred years" (Crosby 2005). Whether this new tactic will suffice to challenge the power of cleaning capital within the new structural context of labour market regulation, though, remains to be seen.

Acknowledgements

The authors wish to acknowledge the assistance of Helen Creed, national president, Liquor, Hospitality and Miscellaneous Workers' Union, Marja Verkerk, of the New Zealand Building Service Contractors, and Michael Crosby, SEIV, Sydney, for providing useful information and papers. We would also like to acknowledge the comments of the two anonymous referees. The usual caveat applies.

Endnotes

[1] It should be recognized that at times of strength unions have generally opposed such compulsory arbitration whilst employers have supported it, particularly when awards have been below increases in the cost of living. By the 1980s, though, employers wished to see it abolished, believing that this would enable them to lower wages.

[2] Arguments relating to high productivity rates tend to be directed at Australia, rather than New Zealand.

[3] This latter figure includes cleaners working in-house and cleaners excluded from the ABS Cleaning Services survey.

[4] One study conducted in response to the privatization of the NSW Government Cleaning Service found that, at the time of privatization, 77% of the cleaners were female, 77% were 40 years of age or older and 42% were from a non-English-speaking backgrounds (Fraser 1997).

References

ABS (Australian Bureau of Statistics) (1990) *Cleaning Services Australia*. Cat No 8672.0 1987–1988

ABS (Australian Bureau of Statistics) (2000) *Cleaning Services Australia*. Cat No 8672.0 1998–1999

ACIRRT (Australian Centre for Industrial Relations Research and Training) (1999) *Australia at Work: Just Managing?* Sydney: Prentice Hall

ACIRRT (Australian Centre for Industrial Relations Research and Training) (2002) *A Comparison of Employment Conditions in Individual Workplace Agreements and Awards in Western Australia*. Report prepared for Commissioner of Workplace Agreements

Aguiar L (2001) Doing cleaning work "scientifically": The reorganization of work in the contract building cleaning industry. *Economic and Industrial Democracy* 22:239–269

AIRC (Australian Industrial Relations Commission) (2005) *Historical Overview*. Canberra: Commonwealth of Australia. http://www.airc.gov.au/about_the_commission/general_information/overview.pdf (last accessed 24 December 2005)

Alcorso C (1991) *Non-English Speaking Background Immigrant Women in the Workforce*. Centre for Multicultural Studies, University of Wollongong and the Office of Multicultural Affairs, Department of Prime Minister and Cabinet

Alcorso C (2002) *Improving Occupational Health and Safety Information to Immigrant Workers in NSW*. Working Paper No 78, ACIRRT. Sydney: University of Sydney

Alexander J (2005) Email correspondence between Shaun Ryan and Statistics New Zealand Information Analyst, Wellington, New Zealand, 19 December

Allen J and Henry N (1996) Fragments of industry and employment: Contract service work and the shift towards precarious employment. In R Crompton, D Gallie and K Purcell (eds) *Changing Forms of Employment* (pp 65–82). London: Routledge

Allen J and Henry N (1997) Ulrick Beck's *Risk Society* at work: Labour and employment in the contract service industries. *Transactions of the Institute of British Geographers* 22:180–197

Australian Senate (2005) *Opposition Senators' Report, Provisions of the Workplace Relations Amendment (Work Choices) Bill 2005, Conduct of the Inquiry, 22 November*. http://www.aph.gov.au/Senate/committee/eet_ctte/wr_workchoices05/report/d01.htm (last accessed 2 December 2005)

Barry M and Wailes N (2004) Contrasting systems? 100 years of arbitration in Australia and New Zealand. *The Journal of Industrial Relations* 46(4):430–447

Barry M and Walsh P (2002) Back to the future? State policies and trade unions in New Zealand. *Proceedings of the Joint Conference of the International Industrial Relations Association (IIRA) 4th Regional Congress of the Americas and the Canadian Industrial Research Association*, Toronto

Beazley K (2005) Speech by Kim Beazley, Leader of the Federal Labor Party and Member of Parliament for Brand, WA, to the House of Representatives, Parliament House, upon the Second Reading of the Work Choices Act, 3 November 2005

Bernstein D (1986) The sub-contracting of cleaning work: A case in the casualisation of labour. *Sociological Review* 34:396–422

BOMA (Building Owners and Managers Association) Four O'Clock Forum (1990) Who's taking who to the cleaners? Or why is cleaning constantly a problem? Sydney

Brosnan P and Walsh P (1998) Employment security in Australia and New Zealand. *Labour and Industry* 8:23–42

Brosnan P and Wilkinson F (1989) *Low Pay and the Minimum Wage*. Wellington: New Zealand Institute of Industrial Relations Research

Buchanan J and Callus R (1993) Efficiency and equity at work: The need for labour market regulation in Australia. *Journal of Industrial Relations* 35(4):515–538

Buchanan J and Watson I (1997) *A Profile of Wage Employees*. Working Paper No 47, ACIRRT. Sydney: University of Sydney

Burgess J and Campbell I (1998) The nature and dimensions of precarious employment in Australia. *Labour and Industry* 8:5–22

Commonwealth of Australia (2005) *Work Choices. A Simpler, Fairer, National Workplace Relations System for Australia*. Canberra: Commonwealth of Australia.

http://www.workchoices.gov.au/ourplan/ourplan/Ourplanforamodernworkplace.htm (last accessed 30 November 2005)

Cooper R (2004) Trade unionism in 2003. *Journal of Industrial Relations* 46(2):213–225

Coyle A (1985) GOING PRIVATE: The implications of privatization for women's work. *Feminist Review* 21:5–23

Creed H (2005) "A Decade of Experience with Workplace Agreements." Unpublished paper. http://www.actu.asn.au/public/papers/helen_creed_paper (last accessed 28 December 2005)

Crosby M (2005) Email correspondence between Andrew Herod and SEIU representative, Sydney, Australia, 9 December and 19 December

Culley M (2002) The cleaner, the waiter, the computer operator: Job change, 1986–2000. *Australian Bulletin of Labour* 38:141–162

Employment Relations Service (2004) *ERA Info* 17 (November)

Fairgo.nsw.gov.au (2005) What will the new workplace laws co$t you. http://www.fairgo.nsw.gov.au/FederalIRChanges/employees/scenarios.html#cleaner (last accessed 7 December 2005)

Fraser L (1997) *Impact of Contracting Out on Female NESB Workers: Case Study of the NSW Government Cleaning Service*. Canberra: Ethnic Communities' Council of NSW Inc

Gold R (1964) In the basement: The apartment building janitor. In P L Berger (ed) *The Human Shape of Work: Studies in the Sociology of Occupation* (pp 1–49). South Bend, IN: Gateway Editions

Hamberger J (1995) *Individual Contracts: Enterprise Bargaining*. Working Paper No 39, ACIRRT. Sydney: University of Sydney

Jeffs R (2001) Australian and New Zealand contract cleaning industry. *European Cleaning Journal* April

Jensen P and Liebenberg B (1995) Government cleaning service. Reforming business in New South Wales. In S Domberger and C Hall (eds) *The Contracting Casebook: Competitive Tendering in Action* (pp 13–31). Canberra: AGPS

Kalb G and Scutella R (2003) *Wage and Employment Rates in New Zealand from 1991 to 2001*. Working Paper No 13/03. Melbourne: Melbourne Institute of Applied Economic and Social Research, University of Melbourne

KiwiCareers (2004) *Salary*. http://www.kiwicareers.govt.nz/jobs/14a_pcu/j64521d.htm#employment (last accessed 15 December 2005)

Leonard M (1992) The modern Cinderellas: Women and the contract cleaning industry in Belfast. In S Arber and N Gilbert (ed) *Women and Working Lives* (pp 148–161). London: Macmillan

LHMU (Liquor, Hospitality and Miscellaneous Workers' Union) (1999) Submission to the Independent Pricing and Regulatory Tribunal, School Cleaning Review, Sydney

LHMU (Liquor, Hospitality and Miscellaneous Workers' Union) (2003) *Confronting the Low Pay Crisis: A New Commitment to Fair Wages and Decent Work*. Submission to the Senate Inquiry into Poverty, Sydney

May R, Walsh P and Otto C (2004) Unions and union membership in New Zealand: Annual review for 2003. *New Zealand Journal of Employment Relations* 29(3):83–96

McLintock A H (ed) (1966) Labour, Department of. In *An Encyclopaedia of New Zealand*. Wellington: Government Printer

Meagher G (2003) *Friend or Flunkey? Paid Domestic Workers in the New Economy*. Sydney: University of New South Wales Press

OECD (Organization for Economic Cooperation and Development) (2002) *OECD in Figures: Statistics on Member Countries*. Paris: Organization for Economic Cooperation and Development

Oxenbridge S (1999) The individualisation of employment relations in New Zealand: Trends and outcomes. In S Deery and R Mitchell (eds) *Employment Relation: Individualisation and Union Exclusion* (pp 227–250). Sydney: Federation Press

Rees G and Fielder S (1992) The services economy, subcontracting and the new employment relations: Contract catering and cleaning. *Work, Employment and Society* 6:345–368

Robinson P (2005) The big clean-out? *The Age* 15 October

Ryan S (2001) "Taken to the cleaners": The peculiarities of employment relations in the NSW contract cleaning industry. Paper presented at 15th AIRAANZ Conference, Wollongong, New South Wales, 31 January–3 February

Sassen S (2001) *The Global City: New York, London, Tokyo*. Princeton, NJ: Princeton University Press

Savage L (1998) Geographies of organizing: Justice for Janitors in Los Angeles. In A Herod (ed) *Organizing the Landscape: Geographical Perspectives on Labor Unionism* (pp 225–252). Minneapolis: University of Minnesota Press

Spotless (2002) *Annual Report*. Melbourne: Spotless

Statistics New Zealand (2000) Labour relations. In *New Zealand Official Yearbook*. Auckland: Government of New Zealand. http://www.stats.govt.nz/quick-facts/people/labour-relations.htm (last accessed 28 November 2005)

Statistics New Zealand (2005) Legislation relevant to Statistics New Zealand. Auckland: Government of New Zealand. http://www.stats.govt.nz/about-us/who-we-are/legislation.htm (last accessed 28 November 2005)

Tempo (1999) After the school bell rings. *intempo* (quarterly magazine for Tempo staff and customers) 1(1):2

Thompson P, Warhurst C and Callaghan G (2001) Ignorant theory and knowledge workers: Interrogating the connections between knowledge, skills and services. *Journal of Management Studies* 38:923–942

Verkerk M (2004) *Cleaning Industry NZ*. Report to Building Service Contractors of New Zealand

Verkerk M (2005) Email correspondence between Shaun Ryan and executive director, NZ Building Service Contractors, 4 August and 19 September

Wailes N 1997 The (re)discovery of the individual employment contract in Australia and New Zealand: The changing demands on industrial relations institutions in a period of economic restructuring. In A Frazer, R McCallum and P Ronfeldt (eds) *Individual Contracts and Workplace Relations* (pp 28–59). Working Paper No 50, ACIRRT. Sydney: University of Sydney

Wailes N and Ramia G (2002) Globalisation, institutions and interests: Comparing recent changes in industrial relations policy in Australia and New Zealand. *Policy, Organisation and Society* 21(1):74–95

Wailes N, Ramia G and Lansbury R D (2003) Interests, institutions and industrial relations. *British Journal of Industrial Relations* 41(4):617–637

Watson I, Buchanan J, Campbell I and Briggs C (2003) *Fragmented Futures: New Challenges in Working Life*. Sydney: Federation Press

Wills J (1998) Space, place and tradition in working-class organization. In A Herod (ed) *Organizing the Landscape: Geographical Perspectives on Labor Unionism* (pp 129–158). University of Minnesota Press: Minneapolis and London

Chapter 5

Manufacturing Modernity: Cleaning, Dirt, and Neoliberalism in Chile

**Patricia Tomic, Ricardo Trumper and
Rodrigo Hidalgo Dattwyler**

The number of lives that enter our own is incalculable.

John Berger (2005:161)

During the 1960s and early 1970s the world witnessed the growth of political movements in a number of Latin American nations dedicated to the redistribution of wealth, resources, and power, and to "putting a human face" on the capitalist economy—if not overthrowing it entirely. The dream, however, did not last long. One by one, the progressive or populist governments of Brazil, Argentina, Uruguay, and Chile were toppled and replaced by repressive military regimes. These regimes soon installed a new political and economic ethic, one of the rationality of the market and what would come to be called neoliberalism. In Chile, as elsewhere, this ethic has had significant impacts upon workers (Winn 2004), spawning a deregulation of labor markets and the erosion of worker protections through the mechanism of, amongst other things, a new labor code (*el Plan Laboral*), introduced in 1979. Significantly, janitors have enjoyed an important, yet contradictory, role in the unfolding of the government's neoliberal plans. On the one hand, as workers on the lower rungs of the socio-economic hierarchy, they have suffered the consequences of neoliberalism more than many, seeing their working conditions worsen as cleaning has increasingly been "professionalized" and outsourced. On the other hand, though, they have been key players rhetorically in a neoliberal project to "modernize" and "civilize" the country, a project which has deliberately conflated ideas of modernity and civilization with those of cleanliness as a way to contrast the "new, modern, neoliberal social order" with the backwardness of underdevelopment.

In this paper, then, we seek to flesh out this conflation of modernity and cleanliness, and to explore its impact upon those workers tasked with ensuring that the spaces of "progress" remain sanitary. We thus focus upon one small sector of the Chilean working classes—janitors—because they are both central to the implementation of the new neoliberal project and because neoliberalism has been quite forcefully imprinted on their lives as they have been obliged to endure the "flexible" work which is characteristic of the contemporary Chilean economic model (Agacino 1995; Echeverría, Solís and Uribe-Echeverría 1998; Escobar 1999; González 1998). Specifically, the paper explores how particular spaces—such as the shopping mall and corporate skyscrapers—have been privileged in neoliberal discourse as representing spaces of modernity and how the practices of maintaining them as clean spaces have been organized. Moreover, we suggest that for the discourse of neoliberal modernization to work, the corridors of mobility—the train system, highways, subways, and streets—which connect such pivotal nodes in the neoliberalizing landscape must also be maintained as sanitary spaces. Put another way, a network of nodes and corridors of asepticism is being carved into the landscape as a way of distinguishing the spaces of modernity from those where the majority of Chileans toil, move, and dwell.

The paper itself is organized as follows. First we summarize the role played by neoliberalism in the economy under both the military dictatorship of General Augusto Pinochet and the civilian governments which have followed since 1990. We then outline how the cleaning industry has been impacted by neoliberalism. Third we explore how the government has marshalled a discourse of cleanliness to buttress support for the "modernization" of Chilean society. Finally, we illustrate how some of the spaces of modernity, such as shopping malls and university campuses, together with the network of transportation routes which connect these geographically, are maintained as clean spaces. Our understanding of just how this is done is based upon data gathered during 25 semi-structured interviews with male and female cleaners in subways, malls, trains, streets, and educational institutions, together with observations from fieldwork conducted in various locations from mid-June to mid-August 2004 and during the months of May and June 2005. The names of the interviewees are fictitious.

Neoliberalism in Chile and the Growing Precariousness of Labor

Prior to 1973, workers had been central actors in Chilean politics and society and, after decades of struggle, had gained some significant economic, social, and political rights (Ruiz Tagle 1985). The military coup of 11 September 1973, however, changed all this. Soon after the Pinochet regime overthrew the government of socialist Salvador Allende, a small group of Chilean economists who were disciples of

Milton Friedman took advantage of the terror clouding the country to lay out a comprehensive neoliberal strategy for the economy. The plan these "Chicago Boys" developed had both economic and political goals (Montecinos 1988; Trumper 1999; Valdés 1989). On the economic front, it intended to bring about the "modernization" of the economy through a number of policy changes designed to wipe away what were perceived to be outmoded forms of ownership and of industrial and commercial organization and operation. For the Friedmanites, then, Chile's future lay in engineering a sharp break with its past by shifting away from the import substitution industrialization (ISI) model which had shaped the Chilean economy since the 1930s and by promoting an export sector based on mining, agriculture, forestry, and fishing (Winn 2004). On the political front, the goal was to punish the working class for supporting a socialist future by eliminating the left-wing political parties and the powerful union movement which had more than a century-old history of organizing and struggle. The ultimate objective, then, was to create an ethics of free markets, to bring about labor's submission to the power of capital, and to encourage the celebration of economic and social individualism in place of the collectivism of the pre-1973 era.

In the process of bringing about the neoliberal transformation of Chile, the period from 1979 to 1981 came to be pivotal (Trumper 1999). By this time the union movement and most opposition to the regime had been effectively crushed and, after years of crisis, the economy had finally started to grow, fuelled by an influx of soft loans and growing exports. At this point, the regime felt sufficiently strong to give new impetus to its neoliberal policies by implementing a crucial piece of legislation, *el Plan Laboral* ("the Labor Code"). Introduced on 2 January 1979, the *Plan* was presented as an important step towards modernizing Chile's economy, which for the *Plan*'s key architect—a young, upper-class economist named José Piñera—meant encouraging "flexibility" in labor markets and weakening workers' rights to such an extent that the trade unions would be made almost irrelevant. Indeed, as it would turn out, the new 1979 Labor Code would eventually end up "eradicat[ing] most of the political leverage the Chilean working classes had won since the 1920s" (Tinsman 2004:262).

The *Plan* itself was fairly simple. On the one hand, it granted workers the right to strike, a right that had been taken away by the dictatorship. On the other, however, it allowed bosses to hire strike-breakers. The result was that even if workers dared to strike—certainly a dangerous proposition under the repressive climate of the times—the strikers would have no legal mechanisms by which to prevent their employers from replacing them. As if that were not enough, the new code also declared that legal strikes could only last up to 60 days. Bosses, meanwhile, could fire workers at will, with or without cause, at the cost of one-month's salary per year worked as severance, but only if the worker had remained in the job for more than six months. The result of this provision was that many employers (particularly those employing unskilled and/or semi-skilled workers who could be quickly trained to do particular tasks) began to engage in a process of constantly turning over their labor force, thereby preventing workers from reaching the end of the six-month trial period and thus qualifying for severance pay. In sum, the *Plan Laboral* increasingly made work sufficiently contingent and insecure as to bring about the flexibility in the labor market which the Chicago Boys felt was necessary to ensure that labor could be deployed most efficiently according to the needs of capital. Arguably, the *precarización del trabajo* ("growing precariousness of work") facilitated by the *Plan* has been no more evident than in the janitorial industry and mirrors the work instability janitors are facing in other countries around the world (see Aguiar; Bezuidenhout and Fakier; Ryan and Herod; and Brody in this collection).[1]

The Making of a Neoliberal Cleaning Industry
Cleaning in Chile has historically been intertwined with gender and class relations in very particular ways and has always been strongly focused upon the home and ideas of domesticity. Whilst domestic cleaning has traditionally been a woman's job, it is also the case that, in the households of people with economic means, cleaning, as well as cooking, laundry, and child rearing, has been historically performed by hired labor. This is reflected today in the fact that there are some 300,000 domestic workers in Chile and domestic workers comprise 13% of all women in the labor force (Centro MICRODATOS 2005). However, it is important to recognize that domestic workers are not just waged laborers and that the social relations between them and their bosses are not simply marked by a monetary exchange. Rather, as in other parts of the world, there are also relations of reciprocity which shape the practise of domestic service (Maher and Staab 2005; Scheper-Hughes 1995). In fact, domestic work in Chile is very much characterized by a kind of *latifundia* social relation—relations of subordination and reciprocity rooted in the rural *hacienda* system that dominated the Chilean countryside until the Agrarian Reform in the 1960s and early 1970s.

Similar conditions order the male-dominated world of non-domestic janitorial work. For example, janitors who work in residential buildings typically fulfill a multiplicity of tasks, such as cleaning, guarding the premises, fixing anything that breaks, opening the entrance doors for residents, carrying residents' groceries, and offering a wide variety of other services to residents—they are, in other words, what Durkheim (1964) would call "generalists". Likewise, until the late 1970s at least, the cleaning of large factories, offices, government institutions, and schools was the preserve of male generalists who were employed directly by the factory, office or institution in which they worked and whose job descriptions typically included conducting errands such as shopping, banking, plumbing, and a multiplicity of other tasks, in addition to cleaning.[2] Although the work of janitors was often devalued, they were nevertheless typically hired on a permanent basis and were integral to an institution's day-to-day operation.

The *Plan Laboral*, however, has led to a radical change in the approach to the cleaning of factories and offices. Thus, whilst during the early days of the Pinochet regime cleaning largely remained firmly anchored in the long-standing practices of the past, by the late 1970s the cleaning tasks performed by janitors started to be outsourced to specialized cleaning companies, such that by 1985 the nature of professional cleaning had been transformed radically.[3] Standards of cleaning also began to change. Specialization and the professionalization of cleaning became central to the industry's efforts to represent itself as a leading symbol of the modernity which the regime wished to engender. Increasingly, cleaning was seen as a task which required the application of "scientific" principles so as to make it more efficient and modern (cf Aguiar 2001), whilst janitors were given the appearance of "professionals" through the adoption of crisp and colorful uniforms. For its part, cleanliness was taken to be a symbol of modernity. Such was the impetus towards professionalization that in 1986 the *Asociación Gremial de Empresas de Limpieza Industrial* (AGEL/ "Association of Industrial Cleaning Companies") would proclaim that "until a few years ago cleaning was normally done by people who [also] performed other functions for the company. The quality of their work was [thus] less than satisfactory. Now, although some companies still do not offer a good service, professionals perform the job [and all] the members of [our association] subscribe to an ethics code" (Rico Enriquez 1986:1; our translation).

However, whilst neoliberal discourses of modernity were beginning to infuse industrial cleaning with ideas of professionalism and specialization in the late 1970s, neoliberal economic policies were plunging the country into one of its worst ever economic crises. High levels of unemployment and poverty, coupled with increasing deficits in housing, health, and sanitation, resulted in social problems of giant proportions, where diseases like louse infections and scabies, the product of poverty, were rampant.[4] The sustained levels of poverty and the crisis of the neoliberal project in the early 1980s eventually resulted in a popular explosion that challenged the military regime and the legitimacy of the dominant classes' project. A spontaneous and massive social movement threatened to marginalize the political parties (Bengoa 1990). In response, the political parties in the right, center, and left negotiated an agreement to quell popular unrest whilst slowly moving towards a civilian regime. At the core of this accord was the strengthening of neoliberalism. Civilians in the Christian Democratic Party, the Socialist Party, and other political organizations thus formed the *Concertación de Partidos por la Democracia* ("Coalition of Parties for Democracy") which eventually took power in 1990 after making a pact with the dictatorship to retain and deepen neoliberalism (Tomic and Trumper 1998).

The outcome of these three decades of neoliberalism in Chile, then, has been the privatization of most economic activities, the promotion of a deregulated and "flexible" labor market relying on contingent work, and the growing penetration of foreign capital into the Chilean economy. This phenomenon is seen in most industries, including professional cleaning and other service industries such as pest control, landscaping, security, and the provision of workers for factories. In the case of cleaning specifically, the moves towards flexibility— particularly through outsourcing—have spawned a significant growth in the number of firms operating in the sector, with the number having increased by 500% since 1980, to the point where there are currently some 470 companies in the cleaning sector in Chile employing almost 31,000 workers (Valenzuela 2005). As foreign capital has penetrated the economy, though, increasing numbers of these firms are being acquired by foreign interests. Perhaps the most emblematic of these acquisitions has been that of Serviman, a Chilean company established in 1979 with operations spanning from Arica in the north to Punta Arenas in the south and which services all types of educational facilities, office buildings, private hospitals, shopping malls, and supermarkets, employing more than 1300 cleaners. In 2004 the company was purchased by the Danish conglomerate ISS, which claims to be the largest professional cleaning company in the world, with 285,000 employees and 125,000 clients in 44 countries. In the same year ISS also purchased Seasur, another Chilean cleaning company with 1600 employees. Such acquisitions are significant, because foreign ownership has strengthened the "flexible" organization of work and has introduced new discourses surrounding such work. For example, ISS contends that its final product is unique because it is based upon "Scandinavian values" of honesty, initiative, responsibility, and high quality services (*El Mercurio* nd)—values which ISS executives presumably believe are lacking amongst Chilean workers. ISS's work practices, though, have also intensified cleaning work. Hence, workers we interviewed in one facility contracted to ISS revealed that the company requires shifts of nine hours a day, with a half-an-hour break for lunch, from Monday to Friday. The cleaners (all women) are paid the minimum wage of 127,000 pesos a month (as of November 2005, around US$245), do not enjoy union protection and work on a short-term contract, which means that at the contract's termination (*término de faena*) ISS is legally entitled to fire them (interview with Juana, Concepción, 15 June 2005), even if the building owner decides to recontract with the company—a practice which contributes significantly to the *precarización del trabajo*.[5]

Modernity and Cleanliness: The Anti-cholera Campaign

Professional cleaning has occupied an important role in efforts to transform Chilean society, for cleanliness and modernity have been strongly linked in the government's neoliberal rhetoric. Certainly, this conflation is not unique to Chile, but it does mean that keeping what are seen to be the spaces of modernity clean has taken on significant rhetorical weight in the discursive reinventing of the Chilean economy as a "modern" one. However, the cholera epidemic which hit Latin America in the early 1990s was also important in focusing public discourse around the notion that the modernization of Chilean society was overdue and that this would be represented by greater cleanliness in both the public and private spheres. Indeed, when cholera broke out after erupting in neighboring Peru and Bolivia, two countries that most Chileans consider less developed than their own, many were shocked. This jolt to national self-esteem would allow the Chilean state to engage in a bio-political strategy (Armstrong 1983) to discipline the population.

The timing of the epidemic, then, proved fortuitous for the new civilian government of Patricio Aylwin, for it presented a discursive "moment" through which it could cast cholera as a disease of social and economic "backwardness" and thus forever link in the public's mind the need for more "modern" political and economic forms, as represented by a resurgent democratization and neoliberalism (to which the administration was ideologically committed), with the need for cleanliness. Hence, although it soon became evident that there was little danger of an actual epidemic in Chile, for good water and sewage treatment facilities protected much of the country's population and the cause of the few cholera cases reported was found to be the consumption of vegetables irrigated with untreated water in rural areas close to Santiago, the government nevertheless used the opportunity to undertake a massive campaign to encourage the population to adopt higher standards of cleanliness (Trumper and Phillips 1997). Indeed, the anti-cholera campaign was structured quite deliberately to push individuals into new and "modern" routines of cleanliness. Specifically, the population—particularly women living in poverty—was made the subject of a forceful campaign that encouraged them to properly wash their hands, clean their toilets, and mop their homes. Posters were placed seemingly everywhere—in the subway system, in poor neighborhoods, and on community billboards. Community meetings were organized in poor neighborhoods to emphasize the importance of washing and disinfection, and supermarkets began posting signs in their vegetable sections stressing cleanliness (Trumper and Phillips 1995:185, italics in original):

Primero lavar [First wash]
Luego preparar [Then cook]

Primero hervir [First boil]
Luego servir [Then serve]

Primero limpiar [First clean]
Luego botar [Then throw out]

Tener buenos hábitos de alimentación e higiene protege su salud
[Good eating and hygiene habits protect your health]

Public areas such as washrooms and restaurants were also the focus of stringent standards of cleanliness as the new asepsis was represented as the embodiment of modernity. In such rhetoric, good hygiene was viewed as a characteristic of "developed" nations, those that had left "backwardness" behind.

Through the anti-cholera campaign, then, modernity and cleanliness became closely enmeshed in the Chilean imagination. In turn, modernity and cleanliness were recursively paired to neoliberalism and the ever-present rationality of the market (Trumper 1999; Trumper and Phillips 1995, 1997). For example, whereas in Chile public toilets had traditionally been unattended but free, as neoliberalism transformed the economy, patrons increasingly had to pay to use them, a development which placed a not inconsiderable burden on the poor, given that public washroom fees can cost the equivalent of a bus fare, a not insignificant sum in the budget of many Chileans.[6] The *quid pro quo* for those who could afford to pay, of course, was that the new fees bought them access to clean, "modern" washrooms, thanks to the poorly paid work of attendants whose job it was to maintain them. However, for such attendants, the marketization of excretion has meant a considerable change in ways of doing business. Thus, Pedro, an attendant/cleaner of a "public" male washroom in Santiago, told us that he had worked as an attendant in the washrooms of a public institution for 25 years. During his first ten years his salary was paid by the institution and was supplemented by some voluntary tips. When standards of cleanliness became more stringent in the early 1990s, however, the institution stopped paying him and he was expected to run the washroom facility on his own, charging fees and keeping whatever money he earned after providing toilet paper, soap, and towels.[7]

Sanitary Landscapes as Spaces of Modernity

If the fear of cholera brought the pursuit of cleanliness to the fore on an individual basis—urging people to wash their hands as a symbol of being "modern" citizens, for instance—then the neoliberal production and management of the broader contemporary built environment has re-enforced cleanliness's importance in the effort to distinguish between the modern and non-modern/"backward" spaces of the country. In this regard, the idea of dirt removal has become intertwined with that of the production more generally of "sanitary" and ordered landscapes. In this section, then, we explore how a number of sites which are considered pivotal to the articulation of a vision of modernity are maintained as clean spaces.

The Shopping Mall

As neoliberalism has proceeded apace, the production of Chilean urban space has increasingly mimicked that of North American business districts (Baudrillard 1988), a phenomenon which is, arguably, no more apparent than in the example of "Sanhattan", a small group of post-modern high-rises in the financial district in a posh area of eastern Santiago whose high glass towers are powerful symbols of progress shining precariously under the grey smog of thousands of cars on the old clogged streets that surround the area. Another central aspect of this mimicking, though, has been the widespread construction of shopping malls, which have brought many commercial activities out of the hustle and bustle of the chaotic *calle* ("street") and into a regulated and climate-controlled arena wherein security and cleanliness can be ensured.[8] There are 11 such malls in Santiago and several strip malls. These facilities, owned by powerful Chilean interests, serve as symbols of Chile's modernity and are very much seen as part of a broader process of transforming particular sectors of the urban landscape into arenas for "modern living". Indeed, they have come to be viewed as the epitome of modernity, with modernity understood as "Americanization" (Cáceres Quiero and Farías Soto 1999).[9]

In the case of Santiago, one of the most exclusive malls is the Alto Las Condes shopping center. Opened in 1993, it is billed by its owners as the "most modern commercial center in South America" (Alto Las Condes 2005).[10] To live up to their image, however, these malls must constantly be kept clean, generally by armies of the very people who are being excluded by Chile's neoliberal "miracle". In such places, dirt is most definitely "out of place" (Douglas 1966), whilst closed-circuit cameras and uniformed security guards complete the picture that such spaces are out-of-bounds for both those things and those people that would sully the spaces of purity and cleanliness which are taken to represent modern Chile. Santiago, however, is not the only place to have experienced the growth of malls as spaces that are representative of modernity and which demand constant vigilance against the intrusions of dirt and all that is associated with it. Similar developments have occurred in places such as Concepción-Talcahuano, a metropolitan area located on the Pacific coast some 500km to the southwest of Santiago and the country's second largest urban conglomerate.

Concepción itself was transformed into a center of heavy industry during Chile's ISI drive, with these industries complementing the already-existing coal mining and light industry in neighboring towns. When the military regime and neoliberalism put an end to import substitution many of the industries in the area closed, for they could not compete with imported goods. The result has been a crumbling infrastructure and poor housing, and growing environmental pressures as fishing and forestry have replaced the dying activities. Finding asepsis is difficult in this place where the humidity of constant rain penetrates the walls of unheated homes and forms puddles on the uneven pavement, and where the smell of fisheries and the plumes of smoke hang in the air. Nevertheless, even as the city's air is almost permanently contaminated with the bouquet of fish and petrochemicals, and even as cockroaches climb the walls of the traditional market, "progress" has arrived in Concepción in the form of the Mall Plaza del Trébol and the practices of professional cleaning by which it is kept sanitized.

Built in 1995 by the Plaza Mall chain half-way between the crumbling port city of Talcahuano and the city of Concepción, the Mall Plaza del Trébol serves as the main space for shopping for Concepción's relatively small middle class. The mall itself is kept clean by janitors employed by EFCO Servicios Generales, a company that has around 900 workers and which also cleans other malls owned by the same conglomerate. The mall cleaners, mostly women, are ever present, dressed in red and blue shirts, with red caps and yellow rubber gloves, and carrying mops and scrapers, as if the company wanted their presence acknowledged as a way of reassuring the mall's customers that they have entered a hygienic space. They mop ceaselessly, persistently purifying the busy space, all the while trying not to get in the customers' way, even as they appear to be invisible to them (cf Aguiar 2005 and Sassen 1991 for similar observations about cleaners in North America).[11] This need to constantly be engaged in the work of cleaning is, perhaps, representative of the place—both literal and metaphorical—where the Mall Plaza del Trébol stands. Thus, sitting in close proximity to one of the most polluted environments in the country and engulfed by the all-pervasive stink of Talcahuano that cannot be avoided once the customers and workers step outside its doors, the mall represents an apparent *cordon sanitaire* of modernity against the backwardness of the surrounding environment where the working classes toil and live. Indeed, such is the power of symbolism here that even the workers appear to have to conform bodily to the demands for a fresh, modern look: no cleaner working in the mall is older than 45 years of age.[12]

Maintaining cleanliness, then, is a central aspect of presenting an image of modernity and progress in places like Santiago's Alto Las Condes mall or the Mall Plaza del Trébol facilities. However, such cleanliness is also sought out frenetically, with different degrees of success, by both state- and privately-owned companies in their old and new buildings, in schools and offices, and even in the country's public transportation system. This latter is particularly significant, because sectors of the public transportation system represent an important network of "corridors" of cleanliness which serve to link the nodal spaces of shopping malls, new skyscrapers, and private schools and universities, corridors through which and in which the middle and upper classes may move, work, play, learn, and live. Such corridors stand in stark contrast to the rest of the country, *el otro Chile* ("the other Chile"), where the great majority of Chileans live their lives and where dirt is decidedly not out of place.

Modernity on Wheels

Trains in Chile have a history that extends back to the mid-nineteenth century. As such, they elicit a sense of nostalgia that other forms of long-distance travel arguably do not. De Certeau (1988), for instance, suggests that there is an unquestionable appeal in train rides, maintaining that the train offers a form of travel that inspires thinking and reflection. Whatever the case, for its part the military dictatorship only treasured patriotic nostalgia and it was not particularly enamored of the country's train system. Consequently, during its tenure, the regime allowed the state-owned train service to deteriorate, eliminating runs and offering a dismal service. Since the 1990s, however, civilian neoliberalism has transformed the capacity for movement into a central tenet of its project to "modernize" the country, seeing the improvement of the nation's transportation infrastructure as essential to the ability of goods (and hence capital) to move from place to place in response to market forces (cf Harvey 1982). Generally, the government shares the ideology of the powerful transportation engineers' lobby (whose preference is for buses and trucks rather than trains) and tends to see modernity symbolized as automobility—that is to say, a system that includes roads, gas stations, and private cars and their drivers (Sheller and Urry 2000). Accordingly, the government has developed close ties with corporations (mostly foreign) in the transportation business who have made considerable investments in a network of paid urban and intercity highways. Automobility as modernity, then, stands in opposition to trains which, for many, symbolize the old Chile. Nevertheless, at the same time, Chile's class reality is such that many people do not own cars, whilst air pollution in urban areas has become a problem of vast proportions, one which sullies the new urban landscape that supposedly reflects the brightness of Chile's neoliberal future.[13]

As a way of providing long-distance mobility options for a market of nostalgic travelers wealthy enough to avoid less comfortable intercity buses, then, in recent years the government has warily reinvested in the state-owned train company, slowly buying second-hand rolling stock from Spain and improving the tracks and rail stations. Certainly, this older infrastructure contrasts greatly with the rising numbers of new imported automobiles and the fleets of state-of-the-art intercity buses, and the train stations look shabby against the large and shiny gas stations located at the privately run concessions which dot the new four-lane modern highways that link the main cities and towns of Chile's urban system. Nonetheless, as a practical matter the government has sought to improve the train system, deciding that if it must have trains then they should at least live up to the image of modernity which it has tried to express with other, more favored, forms of transportation. To do so it has had to overcome one of the leading signs of the deterioration of the train system during the military regime, namely filth: filthy trains, stinking washrooms, and hints of lice and bedbugs in the dormitory cars. Indeed, under the military filth became the symbol of the lack of modernity in train transportation. Today, by way of contrast, the passenger railway system has been dramatically cleaned up in an attempt to match it to the new codes of progress and trains have gained a place in the spaces of modernity, even as they represent a certain ambiguity—the penetration of decrepit and "old fashioned" forms of transportation into a world of post-Fordist, neoliberal modernity. Accordingly, Chileans have increasingly come to see ideas of "modern" cleanliness grafted onto the rolling stock of a bygone era both literally and figuratively as older trains are operated according to "modern" practices of service outsourcing, practices which have served equally to cleanse trains of dirt and to cleanse train employees of the allegedly "primitive" work ethics of the pre-neoliberal age. Thus, as with Chilean society in general, so with the train system has the outsourcing of work to create "flexibility" been widespread—in the case of the state-run train company Empresa de Ferrocarriles del Estado, for instance, all cleaning services are outsourced and workers typically are hired for a *faena*, that is a fixed period of time for which their employers have contracted to offer the services.

Given the penetration of overseas capital into the Chilean economy, much of the cleaning of the railway system is conducted by large, foreign-owned companies—in this case, Cogan Grupo Norte, a company which has its origins in the 2004 purchase of the Chilean cleaning company Cogan (which began operations just after the *Plan Laboral* was initiated) by the Spanish company Grupo Norte. Cogan itself has been associated with intercity transportation since its founding, when it took over cleaning for companies such as the giant Chilean bus conglomerate Turbus. As in the case of shopping malls, cleaners are ever-present on both the trains and in the train stations— on one train on which we conducted observations, a male cleaner in a spiffy blue and red uniform walked through the car eight times in the space of just four hours, collecting papers, bottles, and other garbage, whereas in Santiago's Central Station we counted 22 cleaners one afternoon, men and women in uniform who were constantly sweeping, mopping, picking up the tiniest pieces of paper, and scraping chewing gum off the floors and walls in an attempt to sanitize the place. Such cleaners—particularly the men—also helped to unload passengers and suitcases, offered directions, and served as porters (field notes, Santiago, 1 June 2005). Not all the cleaning in train stations, however, is in the hands of Cogan Grupo Norte, for some of the public washrooms in stations are cleaned by a number of smaller companies who generally specialize in managing toilet facilities. These companies make a profit by selling the right to use the washrooms to passengers and sometimes by offering the possibility of hot showers, with the fee to use the washroom being 150 pesos (US$0.28) and that for a hot shower 2500 pesos (US$4.75). Cleaning, however, is not the only traveler service that has been outsourced. Hence, in Santiago's Central Station a small entrepreneur runs the baggage storage facility and has the franchise for the washroom. The outsourcing of such arrangements for bathrooms is not unusual and is now a constant in most train stations covered by the state-owned train company.

If the train system provides a means for travelers to journey from city to city in a sanitized environment, then the Santiago Metro (subway) system—first opened in 1975 and recently expanded significantly—stands as another powerful indication of the modernization of movement which the neoliberal Chilean state is pursuing, this time as a means to move about the spaces of the city. With thousands of bright lights and walls emblazoned with modern art in selected stations, the Metro is a veritable embodiment of modernity.[14] Although state-owned, it operates very much like a private company, with the work of maintaining its modern, sanitary look contracted out to a transnational cleaning company and conducted by a bevy of highly exploited workers—the Metro's webpage, for instance, proudly indicates that "*Metro de Santiago* is a company with a reduced complement of workers that gets help from outsourcing in . . . security, cleaning, and other tasks" (Metro de Santiago 2005). Mirroring the intercity train stations, uniformed cleaners are readily evident on the Metro's platforms, as are the security guards who protect the stations from "undesirables". Both cleaners and guards are ubiquitous around the ticket booths and platforms, the stairways and corridors, with the cleaners constantly sweeping, mopping, and scraping to make sure that there are no traces of dirt which might represent the penetration of the Metro's orderly, semi-private world by the grime and disorder to be found outside it. Significantly, these workers are uniformed in solid blue coats showing in large letters that they work for Aseos Industriales Casino (field notes, Santiago, 2 June 2005). However, to this name has been added another: Aramark SA. As with many parts of the Chilean economy, then, cleaning here has also been outsourced to a foreign transnational corporation, one that entered the market through buying the local Aseos Industriales Casino and which, despite having worldwide sales of more than US$10 billion and profits of some US$263 million in 2004 (Aramark Corporation 2005), pays its Chilean cleaners a net monthly salary of only 90,000 pesos (US$172) and employs them on contracts lasting just three months.

In terms of the mechanics of keeping the Metro system dirt free, cleaning is a non-stop operation. Hence, two shifts of workers, toiling for eight hours each, keep the subway clean during its hours of operation, whilst a third shift is responsible for waxing the platforms and sanitizing the cars after 11pm, when the trains stop running. Whereas during the day the job of constantly cleaning the platforms is conducted primarily using fairly unobtrusive mops and scrapers, at night these are replaced by industrial vacuum cleaners and floor-polishing machines. The workers we interviewed indicated that, generally, day shifts and night shifts are paid at the same rate but what changes is the gender make-up of the cleaning crews—only men are entitled to operate the heavier or more automated machines, whereas women are usually restricted to mops and dustpans.

Spotless Streets, Modern World

Depending upon where one leaves the Metro, one may or may not encounter modernity and cleanliness. This is because Santiago's urban political fragmentation follows a sharp class divide, wherein five or six wealthy municipalities contrast countless poor ones. The class divide is imprinted in the differential access to, and quality of, medical facilities and schools, in the architectural design and availability of sidewalks, in the number of potholes in the streets, and not least in the amount of garbage piled up in public spaces. Frequently, in Santiago's poorer neighborhoods there is no trash pick-up and the litter collects everywhere. The circuits of modernity through which the middle and upper classes move, by way of contrast, are spotless, for the region's wealthy municipalities spend large sums of money on cleaning so as to keep the appearance of being modern. Given Santiago's urban history, it is in the downtown where the city's two societies—middle- and upper-class "modern" Santiago on the one hand, and *el otro Chile* on the other—most commonly cross paths. The downtown is, then, a decidedly contested space—middle- and upper-class males dominate it during the day whilst the people from poor areas occupy it in the evenings and on weekends. However, as if to keep these latter at bay, the municipality spends vast sums of money on cleaning.

In an effort to give downtown Santiago a modern look, then, the Municipal government and the police try to keep professionals, business people, politicians, and bureaucrats from fleeing the area by keeping it clean and sanitized. Consequently, there is a ubiquitous policing of the downtown's spaces in the form of closed-circuit television cameras, regular uniformed police patrols, plain clothes police officers, and municipal guards dressed in colorful red and black uniforms (field notes, 15 June 2004). Their role is to keep the streets "clean" of street vendors, thieves, political protests, street performers, street theater, dancers, and others who might offend the sensibilities of the middle and upper classes. This form of human cleansing is matched by a furious cleaning of the main streets of the area to remove any form of garbage that might affront accepted norms. Thus, for six days a week, until around six in the evening, the area is constantly kept clean and, therefore, "modern" by a host of relatively inconspicuous cleaners. Only when the powerful and wealthy bureaucrats, professionals, and business people abandon the downtown in the evening and during the weekend does this form of cleaning stop, to be replaced at night by another form of more obtrusive cleaning—the motorized sweepers, the garbage trucks that pick up refuse, and perhaps the most efficient cleaners of all, the throng of independent garbage pickers who rifle through the refuse in search of anything of value, particularly the cardboard which they sell to various recycling companies (field notes, 15 June 2004).

Within this area, arguably the most symbolic street is the *Paseo Ahumada*, a pedestrian walkway that blends postmodern kitsch, fast food chains, bank branches, foreign exchange businesses, and the architecture of a powerful past. Greasy smells, music blaring from loudspeakers competing with elevator music, and the wail of pan-handlers all fill the air. The area itself is kept clean during business hours by the municipality as part of its efforts to enhance an image of safety for the middle and upper classes who occupy it during that period. Through such constant cleaning and surveillance by security guards and video cameras, then, the municipality attempts to assert an image of order and modernity. Significantly, though, the municipality has outsourced cleaning to LIMChile S.A., contract cleaners whose employees, dressed in their blue uniforms and armed with their familiar brooms, mops, scrapers, and cleaning carts, move constantly around and in between pedestrians. This geography of cleanliness is completed by the presence of prominent and modern underground washrooms which, for a fee of 250 pesos (US$0.48), offer spotless facilities with white-uniformed cleaners toiling constantly whilst their colleagues sell the electronic tokens that open the turnstiles to the toilet section and modernity. As with the other workers we interviewed, however, those charged with keeping these corridors of modernity clean typically earn the minimum wage allowed by law.

Of Learning and Cleanliness

The *Paseo Ahumada* has two subway entrances. The *Plaza de Armas* entrance is on line 5 of the subway system. It is as spotless as the other subway stations, complete with an archaeological exhibit. As it runs away from *Plaza de Armas* to the southeast the line passes through poor and decaying manufacturing areas, eventually reaching, 10 stations away, one of the campuses of the *Pontificia Universidad Católica de Chile*, the most prestigious university in the country and one catering primarily to the Chilean elite. This, then, is an island of modernity and cleanliness in a sea of putrefying industrial detritus, an island patrolled by guards dressed in dark, navy blue uniforms (field notes, 17 June 2005) and with a campus kept clean by several contract janitorial companies who bid against each other for the job of cleaning various parts of the campus or even, sometimes, different floors in the same building. One of these companies is INTERSERVICE, which employs some 300 laborers on several campuses of the *Pontificia Universidad Católica*.

INTERSERVICE has recently won a bid to clean several parts of the campus, including sections of particular buildings and some of the grounds surrounding them. The company has about 30 workers who work on this campus, 27 women and three men. The men clean outside, sweeping the streets and sidewalks and picking up the numerous leaves so as to give the place a sense of order and modernity. The women, on the other hand, do the "lighter"—though actually more complex, more dangerous, and dirtier—work inside the buildings, cleaning the tiled floors of the corridors surrounding the interior patios where the classrooms are located, cleaning and tidying up classrooms and washrooms, and emptying the large garbage cans located along the corridors. When it rains or when there is a windstorm, the women may be required to assist their male counterparts in the park areas. In such an environment, cleaning is a perpetual process and workers are expected to keep the place spotless. Thus, the women check the washrooms constantly and between classes tidy up and quickly clean each of the ten classrooms under their care—as one cleaner has stated: "In between classes we have five minutes to pick up garbage, clean the boards, and straighten chairs".[15] Three public washrooms are located at the end of one of the corridors that give access to the classrooms, one for people with disabilities, one for females, and the other for males. Whilst classes are running, the women constantly clean the washrooms, using powerful, industrial-strength chemicals provided by the company. For her part, Lola works the "morning shift", Monday to Friday, from 7am to 4:30pm, with a half-hour break. On Saturdays she works from 8:30am to 3pm. As a morning shift worker she has to make sure that the night shift has left all the classrooms tidy and the washrooms clean. Upon arriving at work, she must open up the classrooms, turn on the lights, and make sure that everything is in place for the new day to begin. The cleaners have no private place in which they can sit during their break from their nine-hour shifts. Meanwhile, INTERSERVICE and the campus authorities abide by the monetary standards of providing purity in the corridors of modernity, and salaries here are not that different from those of the other places studied—120,000 pesos gross, plus a bonus of 5500 pesos and some money for public transportation. Modernity as cleanliness, then, comes at a high price for many.

Conclusion

The concept of cleanliness as an expression of the modernity claimed by neoliberalism for a small part of Chilean society comes at a high human cost, in the form of precarious labor under harsh conditions and low wages. What we have tried to do here is to explore the position of janitors within the Chilean "miracle", at once both subjected to neoliberal working conditions and yet also central to the articulation of the idea that modernity and cleanliness go hand in hand and that if Chile is to maintain the image of a modern society it must also maintain particular standards of cleanliness and hygiene. In so doing, we have explored how the spaces that are seen as symbolic of the new order (malls, offices, and institutions of learning), together with the corridors of mobility which serve to connect them, are maintained as spaces of asepsis. Indeed, discursively the power of the connection between modernity and cleanliness rests, we would argue, on the ability to create such corridors of hygiene which tie together the nodes of the new neoliberal economy and "modern" living into a seamless whole. However, within such a model of economic and social development, cleanliness is not a ubiquitous commodity. Rather, the cleanliness of neoliberal Chile comes at the expense of dirtiness for the majority of the urban areas that lack the means for such perpetual, round-the-clock cleaning. Moreover, such is the nature of this neoliberal project that the areas where those who maintain the image of modernity live are typically inscribed by opposite markers: garbage and pollution, the smell of unkept sewers, and muddy narrow passages or wide roaring high-speed roads from which wealthy Chile passes them by.

Acknowledgments

The authors would like to thank one of the anonymous reviewers for particularly constructive criticisms and Luis Aguiar and Andrew Herod for editorial advice. We also thank Carla Marchant and Rafael Sánchez for their research assistance.

Endnotes

[1] There is no precise English translation of the term "precarización del trabajo". Thus, we translate it here as "growing precariousness of work".

[2] The following piece offers a glimpse into the daily routine of such a janitor:

Vengo del trabajo	I come from work
donde me desempeño	where I officiate
como un humilde	as a humble
auxiliar de aseo	cleaner
de un Instituto de Estudios Superiores	of a post-secondary institution
en la calle Ejército de Santiago	in Ejército Street in Santiago
Y aunque mi sueldo	And although my salary
bordea casi lo miserable	borders on the miserable
No me quejo	I do not complain
Y soy mensajero	And I am a messenger
Y a veces también jardinero	And sometimes also a gardener
Y pago cuentas	And I line up and pay the company bills
Y soy compañero inseparable	And I am an inseparable companion
de la pala y de la escoba	of the broom and the dustpan
Y limpio vidrios	And I clean windows
Y lavo baños	And I wash toilets
Y encero pisos	And I wax floors
Y voy de compras	And I do the shopping
Y hago nudos	And I tie knots

(Mora Ortiz 2001)

[3] It should be pointed out, though, that one such company (SEASIN) had been established as early as December 1973 and had quickly gained cleaning contracts with several large companies.

[4] Following the military coup, workers' living standards steadily declined, such that whereas in 1970 28.5% of the Chilean population was classified as living in poverty, by 1976 the numbers had increased to 56.9% of the population. Poverty levels continued to be high in the 1980s, being at 48.2% in 1984 and, in 1989, the last year of the military regime, remained above 40% (Agacino 1995:129).

[5] Although the cleaners we interviewed here and elsewhere were virtually all earning the minimum wage, some mentioned that they receive extra money for transportation and many earn a bonus at the end of the month (between 5000 and 6000 pesos, or about US$9.50–11.50). The minimum wage, which is set by the Ministry of Finance, was 127,000 pesos per month for 2005 (*El Mercurio* 2005).

[6] One observation we made (field notes, Santiago, 18 June 2005), in a giant bus depot in Santiago, highlights the effects of such marketization: "A woman with a little girl who cannot be more than four years old is at the door of the washrooms, attempting to enter. The attendant requests a fee of 150 pesos and points to the sign that states:

TARIFA ÚNICA
$150
DAMAS-VARONES-NIÑOS
TODOS PAGAN

[FLAT FEE 150 PESOS LADIES–GENTLEMEN–CHILDREN EVERYONE PAYS].

The client answers that she does not have the money, that it is a small girl, that they have been traveling for two hours, that there was no washroom on the bus for such a short trip. The attendant is unmoved. Perhaps she is unmoved because the woman does not have the money and both she and the child are shabbily dressed. The woman turns and leaves whilst the child fidgets and cries 'I must go, I must go'."

[7] Interview with Pedro, Santiago, 12 July 2004.

[8] Beginning in 1990 with the opening of the Mall Plaza Vespucio in the Santiago region, for instance, the Mall Plaza chain has built eight shopping malls in different parts of Chile. The Falabella group, which has a majority share in the ownership of the chain, is one of the larger Chilean conglomerates in the economy.

[9] One tourist guide, for instance, states that "Santiago's districts and the regional capitals [all have] large malls *just like any American shopping mall*" (Gestiones Interculturales 2005, emphasis added).

[10] The Mall Alto Las Condes is part of Cencosud, a Chilean conglomerate controlled by Horst Paulmann, who has vast economic interests in Chile. He started his fortune with a restaurant in the city of La Unión and later built an economic empire. Cencosud has expanded into Argentina, where it also has important investments. Cencosud is building a giant Mall adjacent to the Sanhattan that is due to open in 2006.

[11] We asked at least ten people who visit the mall frequently if they could describe the cleaners' uniforms. Few could in any great detail. This leaves us to wonder why the cleaners are unseen and unnoticed, despite the shiny colors and the significant number of them at work throughout the mall (cf Brody, this issue, who makes a similar observation concerning Thai mall cleaners).

[12] Interview with Juana, Concepción, 15 June 2005.

[13] In Chile car ownership is class-specific. In the poor neighborhoods of southern Santiago only 30% of all families own automobiles, whilst in the wealthy eastern municipalities 80% have cars. Indeed, in the eastern part of Santiago many families own more than one vehicle. Even more telling is that in the east of Santiago, 70% of those who are legally permitted to drive have a driver's license, whilst in other parts of the city the holders of a license decline sharply. Outside the wealthier municipalities, not more than one-third of those 18 years of age and older hold a driver's license (Sectra 2002).

[14] For a collection of photographs of the Metro system, see http://www.skyscrapercity. com/showthread.php?t=195735&page=1&pp=20 (last accessed 24 November 2005).
[15] Interview with Lola, Santiago, 17 June 2005.

References

Agacino R (1995) Todo lo flexible se desvanece: el caso chileno. In R Agacino and M Echeverría (eds) *Flexibilidad y Condiciones de Trabajo Precarias* (pp 105–152). Santiago: PET

Aguiar L L M (2001) Doing cleaning work "scientifically": The reorganization of work in the contract building cleaning industry. *Economic and Industrial Democracy* 22(2):239–269

Aguiar L L M (2005) Cleaners and pop culture representation. *Just Labour* 5:65–79

Alto Las Condes (2005) Quienes somos. http://www.altolascondes.cl (last accessed 25 November 2005)

Aramark Corporation (2005) Company snapshot. http://www.aramark.com/Content Template.aspx?PostingID=369&ChannelID=203 (last accessed 24 November 2005)

Armstrong D (1983) *Political Anatomy of the Body: Medical Knowledge in Britain in the Twentieth Century*. Cambridge: Cambridge University Press

Baudrillard J (1988) *America*. London, New York: Verso

Bengoa J (1990) El tiempo que viene. *Proposiciones* 18:7–14

Berger J (2005) *Here is Where We Meet*. London: Bloomsbury

Cáceres Quiero G and Farías Soto L (1999) El espacio urbano: Efectos de las grandes superficies comerciales en el Santiago de la modernización ininterrumpida 1982–1999. *Ambiente y Desarrollo* XV(4):36–41

Centro MICRODATOS, Departamento de Economía, Universidad de Chile (2005) Encuesta de Ocupación y Desocupación. Comuna de la Pintana, June

De Certeau M (1988) *The Practice of Everyday Life*. Berkeley: University of California Press

Douglas M (1966) *Purity and Danger: An Analysis of the Concepts of Pollution and Taboo*. London: Routledge and Kegan Paul

Durkheim E (1964) *The Division of Labour in Society*. Glencoe, IL: The Free Press

Echeverría M, Solís V and Uribe-Echeverría V (1998) El otro trabajo. Suministro de personas en las empresas. Dirección del Trabajo, Departamento de Estudios, Cuadernos de Investigación 7

El Mercurio (nd) El cuento de hadas de ISS. *El Mercurio* Edición Especiales http://www.edicionesespeciales.elmercurio.com/destacadas/detalle/index.asp?idnoticia=0124052005021X0070049&idcuerpo= (last accessed 2 December 2005)

El Mercurio (2005) Alza de salario mínimo sube gratificación legal. *El Mercurio* en Internet, 7 July. http://www.economiaynegocios.cl/tus_finanzas/tus_finanzas.asp?id=744&numero= (last accessed 2 December 2005)

Escobar P (ed) (1999) *Trabajadores y Empleo en el Chile de los Noventa*. Santiago: Universidad Arcis, LOM, PET

Gestiones Interculturales (2005) Shopping guide. http://www.contactchile.cl/en/chilesantiago-shopping.php (last accessed 26 November 2005)

González C (1998) Notas sobre empleo precario y precarización del empleo en Chile. In *Séptimo Informe Anual de Economía y Trabajo en Chile* (pp 83–136). Santiago: PET

Harvey D (1982) *The Limits to Capital*. Oxford: Basil Blackwell

Maher K H and Staab S (2005) Nanny politics: The dilemmas of working women's empowerment in Santiago, Chile. *International Feminist Journal of Politics* 7(1):71–88

Metro de Santiago (2005) Webpage. http://www.metrosantiago.cl (last accessed 24 November 2005)

Montecinos V (1988) Economics and power: Chilean economists in government, 1958–1985. Unpublished PhD Thesis, University of Pittsburgh

Mora Ortiz U (2001) *Poemas de un Auxiliar de Aseo*. Santiago: Autoedición

Rico Enriquez J (1986) Limpieza, un trabajo profesional. *Asociación de Empresas de Aseo* 2(4):1

Ruiz Tagle J (1985) *El Sindicalismo Chileno después del Plan Laboral*. Santiago: PET

Sassen S (1991) *The Global City: New York, London, Tokyo*. Princeton, NJ: Princeton University Press

Scheper-Hughes N (1995) *Death Without Weeping: The Violence of Everyday Life in Brazil*. Berkeley, CA: University of California Press

SECTRA (Secretaría Interministerial de Planificacion de Transporte) (2002) Encuesta Origen Destino de Viajes 2001 (EOD 2001). Santiago: Gobierno de Chile

Sheller M and Urry J (2000) The city and the car. *International Journal of Urban and Regional Research* 24(4):737–757

Tinsman H (2004) More than victims: Women agricultural workers and social change in rural Chile. In P Winn (ed) *Victims of the Chilean Miracle: Workers and Neoliberalism in the Pinochet Era, 1972–2002* (pp 261–297). Durham, NC: Duke University Press

Tomic P and Trumper R (1998) From a cancerous body to a reconciled family: Legitimizing neoliberalism in Chile. In S Ilcan and L Phillips (eds) *Transgressing Borders: Critical Perspectives on Gender, Household, and Culture* (pp 3–18). Westport, CT: Bergin and Garvey

Trumper R (1999) "Healing" the social body: Silence, terror and re/conciliation in neoliberal Chile *Alternatives: Social Transformation and Humane Governance* 24(1):1–37

Trumper R and Phillips L (1995) Cholera in the time of neo-liberalism: The cases of Chile and Ecuador. *Alternatives: Social Transformation and Humane Governance* 20:165–194

Trumper R and Phillips L (1997) Give me discipline and give me death: Neo-liberalizing health in Chile. *International Journal of Health Services* 27(1):41–55

Valdés J G (1989) *La escuela de Chicago: Operación Chile*. Buenos Aires: Grupo Editorial Z

Valenzuela C (2005) Empresas de aseo industrial han aumentado en 500 por ciento. http://www.agel.cl/actualidad_detalle.asp?Id=15 (last accessed 2 December 2005)

Winn P (ed) (2004) *Victims of the Chilean Miracle: Workers and Neoliberalism in the Pinochet Era, 1973–2002*. Durham, NC: Duke University Press

Section 2

Chapter 6

Introduction: Ethnographies of the Cleaning Body

Andrew Herod and Luis L M Aguiar

In this section of the collection, we shift the analysis to focus upon cleaners' direct experiences of the changing labor processes as brought about by neoliberalism. The ethnographies presented here—of work in malls, hotels, and hospitals—examine how cleaners' bodies, as instruments of labor, are forced to adapt to an accelerating work pace and new disciplinary regimes as managers seek to reorganize the labor process. Of course, in and of itself this is not new, for the capitalist labor process has always molded bodies to its demands for the creation of surplus value, even as it has often produced "deformities, pathologies, [and] sickness" (Harvey 2000:103) in those very bodies. However, whereas throughout most of the 20th century—at least in many of the nation-states of the global North—the state had a significant responsibility for maintaining and rehabilitating the sick body, today the state is reneging on this responsibility and appealing to "individual responsibility" (such as through calling for the voluntary application of ergonomic standards and denying workers access to healthcare if they, for example, drink alcohol or use tobacco products) to remedy the bodily sicknesses caused by the economic system within which workers must toil. In the new social Darwinist economic environment of neoliberalism (Giroux 2004), then, workers' bodies must become sufficiently flexible and adept as to withstand the new work pace and regimes of discipline (Leslie and Butz 1998:376), whilst the nature of contemporary employment relations—the shift to the use of temporary and part-time workers on short-term contracts—means that those workers who cannot become so or whose bodies become worn out through their labors are replaced, disposed of, and left to their own devices.

Certainly, the manner of work reorganization in the cleaning workplace differs in many ways from models of "lean production" in some other work realms, such as in automobile manufacturing (Aguiar 2001). Nonetheless, there are parallels. Thus, as Barbara Ehrenreich (2003:96) shows, global cleaning companies are laying down rules on how to clean which are "enforc[ing] a factorylike—or even convent-like—discipline on their far-flung employees". What is especially significant, however, is that in this process of cleaning work reorgan-ization, bodies are being re-inscribed in racial and ethnicized terms to "naturally" fit the presumed association between specific work assignments and particular morphological types (Waldinger and Lichter 2003:8)—for instance, in the global North it is frequently immigrants from the global South who fill such positions, whereas in the global South processes of rapid urbanization and projects of modernization often suck in rural migrants (many of whom are ethnic or linguistic minorities in their own countries) to work cleaning jobs. Un-inscribing the racialization of bodies in the cleaning industry, then, is one of the key challenges for progressives focused on making changes therein. This is especially so given that such racialization practices frequently overlie other divisions, such as those between "core" and "peripheral" workers in the same workplace which have developed as managerial strategies have institutionalized a division between those workers whom management deems necessary to keep the workplace functioning and those the need for whom changes according to the shifts in the business cycle. Hence, it is often the case (in the global North, at least) that white workers have come to serve as the "core" cleaning workforce whereas visible minorities increasingly make up the "peripheral" workforce, a division which has important consequences for workers' solidarity and the ability of unions to develop cohesive approaches to workplace issues.

It is within this context of how bodies are disciplined in time and space, then, that the three papers in this section examine the treat-ment of the cleaning body in the contemporary workplace. In the first paper, Alyson Brody draws on her study of Laotian cleaners in a Thai shopping mall to show how, in Thailand, malls are seen as ex-pressions of "progress" in a project of modernity and nation-building, spaces of consumption for the new, "modern", urban middle classes. Significantly, in this project cleaners play a central role, maintaining the spaces of the mall as sanitary as a way of reinforcing longstanding associations between cleanliness and civilization. However, in order to maintain the illusion of the mall as a safe and clean space for the public display of middle classness (Simon 2004), such cleaners—who are culturally marked as "backwards" and "uncivilized" as a result of coming from the poorest rural area in Thailand—must be made invis-ible to the mall's patrons. Brody, however, shows that such discourses of "invisibility", "docility", and "backwardness" are not as totalizing as they appear. Hence, cleaners take advantage of the fissures in these discourses to create their own spaces of resistance by reinscribing cleaning work as "fun" and the mall as a space wherein they may engage in flirtatious and other behaviors.

The second paper in this section—by Ana María Seifert and Karen Messing—shifts from the ethnography of the shopping mall to the "back of the house" (Zukin 1995) of two hotels in Montréal, the corridors and rooms where cleaners toil according to a new "flexibile" model of work intensification and outsourcing. As Seifert and Messing explain, this model is being imposed by an aggressive hotel managerial agenda focusing on expanding market share and gaining greater profit for shareholders. Such a model manifests itself in cleaners' worklife through the introduction of, amongst others things, more items to clean in individual rooms (eg coffeemakers, bars, baskets, and trays), new decorations such as mirrors (which are often hard for cleaners to reach), and heavier and larger mattresses, all of which have resulted in an intensification of the physical effort and agility required by cleaners to perform their jobs and produced negative physiological and psychological impacts for them. Hence, for instance, new work regimes are complicating cleaners' ability to organize their lives outside the workplace as many must now work "on-call" as "just-in-time" workers, thereby forgoing the possibility of earning income elsewhere and forcing them to constantly rearrange child care and other activities. Equally, the outsourcing of some hotel activities (eg laundry) has resulted in more work for cleaners because of the poor quality of the linen returned to the hotel, an outcome which delays cleaners' ability to complete their rooms in the allotted time. In response, cleaners have invented various strategies to cope with increasing workloads, including beginning to change items in the rooms even before the current occupant leaves and distributing cleaning tasks over several days in order to have less to do when guests finally depart. Such strategies, however, are generally available only to full-time cleaners who have fixed shifts and room assignments, a factor which is slowly driving a wedge between the "regular" and the just-in-time workers (often, visible minorities) in the hotels.

In the final paper in this section Karen Søgaard, Anne Katrine Blangsted, Andrew Herod, and Lotte Finsen examine the stresses (particularly cardiovascular and musculoskeletal) placed on the cleaning body in a number of Danish hospitals as a result of repetitive work and its intensification. Specifically, they question how the labor process might be organized differently so as to allay some of these bodily stresses. Drawing upon research conducted as part of a European Union-sponsored investigation of how task variability might be used to improve cleaners' worklife, the authors conclude that simply introducing variation into how individual cleaning tasks are performed—as in switching between mopping in a banana-like fashion and in a figure-eight rotation—does little to reduce the physical stress on the cleaning body. Instead, they suggest, a more substantive variation in cleaners' workday tasks—such as job enlargement, whereby cleaners combine job variation by performing cleaning and non-cleaning tasks—is a more effective means of reducing stress on the body. However, whereas such job enlargement has the potential to reduce bodily damage by providing a different stress profile on the body, the turn towards increased outsourcing and work intensification makes it increasingly difficult to implement such a strategy in the contemporary workplace. Put another way, as neoliberal work practices become more widespread, the bodily impacts of these practices become more extensive. Moreover, as the authors point out, it is, in fact, possible to measure these corporeal impacts empirically.

References

Aguiar L L M (2001) Doing cleaning work "scientifically": The reorganization of work in the contract building cleaning industry. *Economic and Industrial Democracy* 22(2):239–269

Ehrenreich B (2003) Maid to order. In B Ehrenreich and AR Hochschild (eds) *Global Woman* (pp 85–103). New York: Metropolitan Books

Giroux H (2004) *The Terror of Neoliberalism*. Aurora, Ontario: Garamond Press

Harvey D (2000) *Spaces of Hope*. Berkeley: University of California Press

Leslie D and Butz D (1998) "GM suicide": Flexibility, space and the injured body. *Economic Geography* 74(4):360–378

Simon B (2004) *Boardwalk of Dreams*. New York: Oxford University Press

Waldinger R and Lichter M I (2003) *How the Other Half Works*. Berkeley: University of California

Zukin S (1995) *The Cultures of Cities*. Malden, MA: Blackwell

Chapter 7

The Cleaners You Aren't Meant to See: Order, Hygiene and Everyday Politics in a Bangkok Shopping Mall

Alyson Brody

Introduction

Thailand is a country renowned for its beautiful temples, beautiful beaches, and beautiful women. Few would add the humble shopping mall to this list, yet over the past decade these elaborate buildings have started to redefine Thai urban space, and spending time shopping and eating in them has become an intrinsic part of the modern urban experience, especially for the Bangkok middle classes. In an overcrowded city with little in the way of public parks and other venues for recreation, shopping malls have become "the predominant spaces for leisure and recreation in the city" (Wilson 2004:106). They are the new temples, with their ornate designs and promise of a better, more fulfilled life. Significantly, in such spaces of modernity dirt of any kind is very much viewed as "matter out of place" (Douglas 1966), an interloper sullying the glittering perfection of modern Thai life. Given this perception, then, this paper is about the people who manage the removal of dirt in Bangkok shopping malls—the cleaners who are not part of the idealized world the malls are selling but who are, nonetheless, essential to its creation and maintenance.

Despite the fact that the janitorial industry employs thousands of rural migrants across urban Thailand, it is perhaps telling that most research on Thai migration has focused on prostitution and, to a lesser extent, factory employees (see, for example, Boonchalaksi and Guest 1994; Mills 1993, 1997, 1999; Odzer 1994; Phongpaichit 1982; Thitsa 1980).[1] Meanwhile, the less sensational topic of cleaning has received virtually no attention. This is somewhat curious, for combating dirt has become central to many Thai visions of what it means to talk about the development of a "modern" way of life. As part of an effort both to rectify somewhat this empirical lacuna and to explore aspects of contemporary Thai nation building, then, in this paper I will explore the significance of the apparently mundane topic of cleaning for the way in which it provides a window onto deeper understandings of concepts of "Thai Development" and of the "invisible", yet essential, people who sustain the modern Thai economy.

The paper is organized into three main sections. The first sets the notion of cleanliness within the context of contemporary Thai Development, discussing how, by marking the boundaries between what is considered civilized and what is not, perceptions of hygiene and dirt have become metaphors for a wider concept of "Progress". It historicizes these issues, then considers their relevance in the context of a large, central Bangkok shopping mall with a middle-class clientele. Further, it discusses the tacit yet central role of cleanliness in maintaining the mall's seamless, modern, consumer-oriented image. The second section of the paper provides an ethnography of some of the women migrants, largely those from rural northeastern Thailand, who constitute the workforce of cleaners employed to maintain these spaces. In presenting this ethnography I take the workplace of the shopping mall to be a specific site for the operation of particular mechanisms of power. Primarily, I discuss cleaners' work environment, focusing on how notions of civilization and progress are translated through material effects which attempt to produce and control cleaners' bodies in space and time. Specifically, I note how uniformity, docility, hierarchies, and cleanliness play an important role in constructing idealized representations of the "Modern Thai workforce", and consider how the regulation of time, space, and bodies is seen as being central to transforming "undisciplined" rural workers into "disciplined" modern urbanites. The final section is concerned with the ways in which the cleaners position themselves within these discourses. In particular, I present ways in which they seek to reclaim themselves against the threat of uniformity and control. My main focus is on indirect forms of agency forged through everyday strategies in and around the workplace. These strategies fall into the category of "everyday resistance" (Scott 1985), forms of resistance which ultimately may have more collective significance than do explicit protests (Turton 1984:65).

In highlighting aspects of these cleaners' lives, the work presented here seeks to respond to Escobar's (1995:187) criticism of the common assumption in scholarship concerning development and human rights that "a 'survival strategy' cannot simultaneously be a political strategy". Like Constable (1997) in her analysis of Filipina maids in Hong Kong, I am reluctant to suggest that the cleaners are *either* resistant *or* they are subordinate. Rather, as explained in Bourdieu's (1977) concept of "habitus", the cleaners I studied outwardly conformed to prescribed codes of practice but sought to claim what I refer to as "small freedoms" within these practices. Such freedoms may appear unremarkable in a broader vocabulary of resistance but they were, I argue, meaningful in the overall nexus of labor and culture within which the women operated. As I will show, the pursuit of these freedoms was significant at an immediate level. But, at a higher level, their pursuit also provided a subtle critique of Bangkok-centric notions of Progress. Concurrently, there was awareness among the women that they could only push the boundaries so far, since economic disparities between rural and urban areas and their own poverty and lack of education made them reliant upon the extant relations of power within which they found themselves for their continued economic survival.

Marking the Boundaries of the Civilized: Between Dirt and Development

Background

In early 2005, Thaksin Shinawatra was re-elected Thai premier. This was a momentous occasion, marking the first time a Thai Prime Minister had secured a second consecutive term in office. The public, however, is quite divided in its support for Thaksin. Although his particular brand of neoliberal politics (sometimes called "Thaksinomics") and his obvious global business ambitions have begun to boost an economy that was close to collapse in 1997, his policies have also left large sectors of the population, notably those engaged in rural livelihoods, less than satisfied. Whilst neoliberalism may have created some wealth for those at the top of the Thai socio-economic ladder, concentrated in Bangkok, there has been little "trickle down" to poorer people in the rural hinterland and/or urban slums. As a result, the gap between rich and poor has become increasingly wide over the past decade (see Phongpaichit and Baker 1995; Ungpakorn 1999). As the urban–rural divide in particular has grown, huge numbers of rural migrants have flooded into Bangkok to take advantage of whatever employment opportunities may exist there (Fuller et al 1983; Fuller and Lightfoot 1984; Hirsch 1990; Phongpaichit and Baker 1995; Thorbek 1987). Unable to survive on farming alone in an increasingly cash-driven economy, these migrants are obliged to seek work in the very place that has been built on the policies that exclude them.

Although they are usually represented as a necessary response to the 1997 collapse of the Thai baht, significantly Thaksin's policies are actually the most recent incarnation of a larger plan of development for the country which stretches back over four decades and which has transformed rural Thailand (see, for example, Office of the Prime

Minister 1996, 2001). Thus, beginning in the 1950s, the Thai government started investing heavily in export-oriented agricultural activities, particularly rice. This development was initially supported by the World Bank and other foreign loans and structured around a series of five-year plans. These plans marked Thailand's transition from a predominantly insular, rural economy to a global player as farmers were encouraged to concentrate their efforts on production for export, efforts which earned them the accolade of being the "backbone" of the nation's economy. However, whereas early development was based upon state support for agriculture, during the 1980s the government began withdrawing subsidies and credit facilities for small-scale farmers as low global prices for agricultural exports, the result largely of fierce competition from countries like China, precipitated a shift in economic emphasis away from agriculture and towards promoting for export the manufacture of goods such as canned seafood, computers and their parts (Thailand's leading export earner for some twenty years), toys, cosmetics, auto parts, electronics, and a host of other manufactured goods (see Phongpaichit and Baker 1995). Such manufacturing, largely based in and around Bangkok, fueled a rapid growth of the Thai economy, such that from 1985 to 1995 Thailand enjoyed the world's fastest rate of economic expansion (almost 9% annually).

Perhaps not surprisingly, as the Thai government withdrew support for small-scale farmers, many began to migrate to urban areas—particularly Bangkok—in search of better-paid opportunities in industry or services. According to Phongpaichit and Baker (1995:153), over two million people came to Bangkok from rural areas in the 1980s and "possibly a further million entered the workforce on a temporary basis". Of these migrants, approximately half are from Isan in northeast Thailand, a region which is the country's poorest and largely dependent on agriculture but is beset by both yearly droughts and floods, and which suffers poor soil conditions. The majority of people from this region are of Lao ethnic origin and are considered by many Thais—particularly richer, urbanized Thais—to be hard-working and honest, if somewhat provincial. Women account for almost half these migrants, both because of companies' preference for young, female employees who are cheaper to employ and considered more responsible and obedient than are men (Chouwilai 2000:85), and because the breakdown of traditional social roles in rural areas (the result of the agrarian transformation begun in the 1950s) has allowed many young women to leave the land and come to Bangkok to satisfy their desire to participate in "the sophistication and modernity of [the] urban life" (Mills 1993:10) that they have glimpsed on television.

Vast numbers of such women are now employed as janitors. However, although they may escape many of the constraints of traditional rural life to enjoy the freedoms offered by Bangkok, such as access to cheap fashions and other up-to-date consumer goods, many such women still continue to play the role of "dutiful daughter", sending money home to their parents and contributing to local Buddhist ceremonies.[2] Whilst the cost of living in Bangkok means that these women often cannot adequately fulfill either role (liberated, modern woman or dutiful, traditional daughter), many nonetheless continue to migrate to the city because, in doing so, they can translate their experiences and earnings into forms of social capital that, though derived from a "modern", urban context, are nevertheless meaningful in a rural one—in the eyes of their contemporaries back home they are seen to have "succeeded" (Brody 2003).

Thai Development, Civilization, and Cleanliness

Whilst economic growth is seen as an end in and of itself, capital "D" Development in Thailand has also been linked to a higher goal—that of "civilizing" the nation (Barmé 1993; Reynolds 1991). This linkage has a long history in the country. Thus, the respected and progressive King Mongkut and his son, Chulalongkorn, are widely credited with having brought modern, Western ideas and aspects of culture to Thailand during the late nineteenth and early twentieth centuries (Turton 2000). Such ideas took material form through the building of roads, the signing of trade agreements, and the introduction of Western-style education and dress for the privileged. Yet, perhaps, just as pervasive as these material changes was the inculcation of a number of concepts that defined and shaped national perceptions of these processes. Of these, an overriding concern was that of "*khwam siwilai*" ("civilization", "being civilized"), which was grounded in Western theories of culture and social evolution, and which continues to inform contemporary Thai understandings of "progress". After 1932, when the monarchy was overthrown and a military government took power, these ideas underpinned broader ambitions of unification for a country that was regionally fragmented and whose people did not think of themselves as being part of a single nation. Hence, along with the name change from Siam to Thailand in 1938, the government embarked on a deliberate policy to instill a sense of nationhood or "Thai-ness" into the populace. Significantly, a central theme to emerge in the subsequent proliferation of discourses about Thailand— its people, culture, and society—was that of the civilizing processes the nation was undergoing.

The influential writer Rajadhon, whose work is still widely read in schools and other institutions, was among those who reinforced perceptions of civilization as a process in which societies move from a simple to a complex state. He explained that civilization is based on rules for behavior, especially "qualities of neatness and order"

(1972:14). Such explanations had two fundamental effects: first, they reinforced the message that Thailand was a unified nation of people on a common journey towards the ultimate goal of civilization; and second, they placed Bangkok at the leading edge of this journey, such that rural areas were seen to be lagging behind and in need of education in proper manners, proper ways of speaking and, above all, cleanliness and order. Isan, in particular, was placed at the tail-end of this trajectory, with its Lao customs and language being linked to the Lao nation, patronizingly referred to as Thailand's "little brother", an epithet bound up in old rivalries and a brief period of colonization by the Siamese over the Lao kingdom in the mid-seventeenth century (Evans 1999; Ngaosyvathn and Ngaosyvathn 1994).

Since the 1950s, these notions of progress have been distilled into the idea of development (*gan pattana*), one aspect of which has been official programs of improvement for villages (see Hirsch 1990; Kemp 1988). Such programs have focused not only on improving basic infrastructure, but also they have aimed to influence behavior through the disciplining of bodies. Thus, Boonmathya's (1997:136) research amongst villagers who have participated in such programs revealed that, in one northeastern village, "Government officials co-operated with the village school teachers to train children to speak politely by using the Central Thai language and [to greet] the elders in a polite manner by bowing . . . their torso and heads as low as possible and walk[ing] lightly". These policies and discourses serve to frame rural Thailand in ways that echo Pigg's (1992:507) observations about the generic Nepalese village that has been constructed ideologically through layers of overlapping, mutually reinforcing narrative, a process which represents "the village" as a "space of backwardness". Such a representation enables urban dwellers to both distance themselves from their "uncivilized" country cousins and to embrace them as active participants in their nation's quest for development. Hence, as Pigg points out, such discussions about rural Thailand, its villages and people, make it "knowable" and thus "disciplinable" even to those who have never set foot in a rural village.

Associations between backwardness, rurality, and dirt, then, are an important part of the Thai development and modernization discourse. Certainly, many cultures have seen dirt as marking the boundary between order and disorder, the civilized and the uncivilized, and, therefore, between purity and potential danger, a representation which makes its elimination imperative to a sense of control over natural forces (Douglas 1966). Hoy (1995), for example, historicizes the pursuit of hygiene in the United States, linking it to a process of Americanization for immigrants from Europe and, especially, Africa, who were seen as dirty and in need of education about hygiene. Tomic et al (this collection) illustrate how the Chilean military used the erasure of dirt as a metaphor for the erasure of what it saw as "backward" economic systems, a metaphor continued by the present neoliberal civilian government. In the case of Thailand, the notion of hygiene carries enormous symbolic significance and is perceived as marking an important difference between "developed", "civilized" (*siwilai*) places and "backward", "uncivilized" (*la samai*) places. Indeed, the degree to which this idea of cleanliness holds sway over the popular imagination has been illustrated recently in a furor involving a young female Thai singer who referred to Lao women as "dirty" during a television broadcast in 2000, a broadcast which was also shown in Laos. This declaration resulted in an outcry from the Lao women's union and Lao embassy officials (*The Nation* 2000). It was clear, however, that the tumult had not simply been caused by the young singer's naïve comments but, rather, was a reaction to deeper prejudices about the relatively undeveloped status of Laos compared with Thailand, prejudices in which dirt is part of the larger discourse concerning levels of development within Thailand, particularly between the predominantly ethnic Lao Thais in rural Isan and the urban middle classes in places such as Bangkok.

From a personal point of view, these discourses were made obvious to me in conversations with middle-class urban Thais who were always concerned when I spent time in the Thai countryside that it would be "very dirty". In the majority of cases, such observations were based upon ideas about "the Isan village" rather than upon direct experience of such places. Certainly, such discourses of dirt may be rooted in perceptions that Isan people have little spare water for washing or cleaning their houses during the long droughts to which they are often subjected. But, in my experience, even though the villages could be muddy during the rainy season, house interiors were always spotless and people showered at least once a day. Rather, it seems, the allegory stands as a marker of difference, the construction of an "Other" against which "modern" Thais can measure their own advancements towards "civilization". Given the significance of dirt as a symbol of class difference and for ideas of "the modern", then, its visible management is very much key to the projection of a certain degree of modernity in the space of the shopping mall—itself seen as an embodiment of modern life—and underpins a comprehensive system of maintaining discipline over bodies and space.

The Shopping Mall, Modern Culture, and an Overview of the Maintainers of Cleanliness

David Harvey has noted the significance of space in shaping contemporary urban experiences. He makes the point that control over space is the prerogative of the powerful and that the significance of space

lies in the possibilities of communicating and reinforcing social hierarchies and authority through "spatial organization and symbolism"
(1989:186). Harvey's comments are useful in finding a perspective
on the city of Bangkok, where uses of space, struggles over space,
and lack of space are implicit features of urbanization.[3] The countless newly erected, modern "glittering towers" convey messages of
affluence and progress. A defining space of the contemporary urban
experience in Bangkok is the shopping mall, a place where ideal
environments can be carved out, away from the overpowering heat,
crowds, and traffic fumes of the wider city. Here, a cornucopia of
consumption possibilities is available for perusal, from designer
clothes to imported foods. The mall satisfies a national obsession with
shopping, which until recently was largely carried out in open markets
and Chinese-owned shops in formal and informal spaces of towns and
cities, whilst the promise of air-conditioned comfort adds the patina of
luxury that modernity represents to many.[4] As Wilson (2004) points
out, however, malls are more than just shopping centers. In the space-
restricted, hot, polluted city, they serve as all-encompassing leisure
spaces, places where families and young couples alike can spend the
day eating in restaurants, drinking coffee, seeing movies, and even
go ice-skating or bowling. They are, in other words, hyper-modern
expressions of new Thai middle-class lifestyles and mentalities (cf
Tomic, Trumper and Hidalgo Dattwyler, this collection).

In order for the associations between cleanliness, civilization, and
modernity to be kept alive, though, these public spaces need to be
constantly maintained and, above all, kept spotless. Given the highly
evocative conceptions of dirt and cleanliness outlined above, it is not
surprising that the employment of an army of cleaners is an integral
aspect of urban shopping mall culture. The ongoing task of hygiene
maintenance, however perfunctory, helps to create and sustain the
illusion and image of order around which urban "civilization" and
wealth are predicated. That the bodies of poor, "uncivilized" and
"dirty" rural women and men are used for the task—even as they are
manipulated so as to be largely invisible to the consuming middle-
class public—is, perhaps, the greatest irony here, though it is no
accident that such rural migrants' labor is considered by many middle-
class urban dwellers to be a means to survey and correct their sloth,
ignorance, and sloppiness (cf Foucault 1975), a theme to which I shall
return below. Thus, demonstrating the capacity to employ others to
do this work, usually for low wages, is not only a visual signal of
prosperity in the public arena of a shopping mall but it is also an
integral aspect of middle-class domesticity in Thailand and across
much of Southeast Asia, where maids are often employed in private
houses—Chin (1998:12), for example, notes a reliance upon maids as
a central aspect of the contemporary Malay middle-class lifestyle that

revolves around the conspicuous consumption of certain goods and services as a key way to construct identities and lifestyles which distinguish the nascent middle class from the lower or working classes.

With regard to my own investigation of these workers' positions as maintainers of a physically clean shopping mall—and hence as bulwarks against the incursion of backwardness and unrefined ways into such spaces of modernity—my point of entry to conducting fieldwork was through the formal channels of the managers of a cleaning company which supplied maintenance staff for a large, well-known shopping mall in the center of Bangkok, as well as for other businesses. The market for building maintenance has been cornered by a few prominent companies whose professionalism and modernity is reflected in their choice of English names, such as "PCS" (Property Cleaning Services). Most of the companies are advertised in a global directory, particularly those who have earned the international standards mark ISO 9000 2000, and they compete for contracts with the many international businesses whose presence marks Thailand's entry into the global economy.[5] To achieve this degree of internationalism, Western-educated, English-speaking managers are required (the head of the cleaning company that serviced the particular mall in which I was conducting my research, for instance, was educated in the United States). This internationalism extends to the potential of viewing cleaning companies as an investment opportunity, and at least one company was encouraging the purchase of its shares over the Internet.

The majority of cleaners in the mall were women. Although there were some male cleaners, who mainly seemed to be occupied with manual tasks such as fixing lights, the majority of men employed by the company were security guards. Their role was mainly to prevent "undesirables" from entering the mall and spoiling its exclusive environment. Although I passively observed the men's behavior and attitudes and recorded the women's opinions of them, it was much easier to form close bonds with the women. In addition, my research intentions were to focus specifically on women's practices and perceptions. Accordingly, in this paper I concentrate mainly on the women at work, as I observed them. I also spent time with the women outside work, in and around the extensive slum community where they lived, which was located next to the mall but was hidden from public view by iron fences.

For its part, the cleaning and maintenance company that employed the cleaners had an office in the basement of the mall but its operations were overseen and monitored by the mall administration. During the period of my research, 150 cleaners were employed at the mall, of whom only 20 were men. The workers ranged in age from 18 to 57 years old, although most of the women were between 25 and 40, and the majority were aged above 30, a sharp contrast with factory workers, who tend to be below 30. One reason for this age difference might be that factories tend to "retire" women when they reach 28 because they become too expensive to hire and the expectation is that they will marry or move to other employment (Chouwilai 2000). Working patterns at the mall were varied, and the work was not contracted but was calculated on a daily wage. This lack of a formal agreement meant that the women could take time off, sometimes for extended periods, to visit their home villages and participate in planting or harvesting or spend time with their children, who were often left in the care of relatives. This flexibility seemed to suit many of the women but it also meant that they were potentially in a less secure position with regard to their work and perhaps felt less able to complain about particular issues. The transience of the workforce also meant that it was difficult for unions to develop the kinds of long-term relationships with workers which are often necessary for successful organizing, especially against a background of resistance from employers (see Brody 2000). Despite the lack of a contract, though, the women did have certain rights. Thus, the company honored recent Thai legislation allowing the cleaners to claim six days' paid leave and three months' maternity leave on minimum wage after having worked three months for the company. The longest period spent working at the mall was five years and the shortest was one month. The basic wage was 162 baht per day (approximately US$5 at the time of research), with increments of 20 baht after one year of work. The cleaners worked from 8am to 8:30pm, six days a week. Although they were expected to take a day off, many cleaners would work overtime for the extra money.

Modes of Discipline: Producing Docile Bodies, Marking Space

In this second part of the paper I provide an ethnography of the cleaners employed at the mall. In discussing the various technologies that combined to create a culture of uniformity, obedience, and hierarchy within the shopping mall I start with an analysis of spaces within it, that is to say the divisions in physical and temporal space, and the separations between people. I also explore gendered work relations within the mall, together with the ways in which these were reinforced. Finally, I consider deliberate strategies aimed at imposing "urban knowledge" and values on the cleaners.

Place, Space, and Hierarchies in the Mall

The organization of space can be the most eloquent expression of the social relations occurring within that space. Thus, as Foucault

(1975:143) has noted, the supervision of workers in large-scale industry relies on their organization and deliberate placing within work areas as a means of maintaining discipline and avoiding the potentially disruptive elements of disorder:

> Each individual has his own place, and each place its individual. Avoid distribution in groups; break up collective dispositions; analyze confused, massive or transient pluralities. Disciplined space tends to be divided into as many sections as there are bodies or elements to be distributed. One must eliminate the effects of imprecise distributions, the uncontrolled disappearance of individuals, their diffuse circulation . . . [The] aim is to establish presences and absences, to know where and how to locate individuals . . . to be able at each moment to supervise the conduct of each individual . . . it is a procedure . . . aimed at knowing, mastering and using.

The assignment of a "place" for workers' bodies was both a physical and a social exercise within the mall. Control over movements in space was a strategic management tool, obviating the potential for idle gossip or wandering beyond the range of supervision. For example, those who cleaned the toilet areas would have to come into work an hour before the others two or three days a week in order to thoroughly scour the floors and toilet cubicles, ready for inspection. They could not move from their assigned restrooms, except during breaks. Those in the main body of the mall kept to a specific area but were chastised if they were not seen to be working constantly. If I stopped to talk to any women who were having a short rest, the arrival of a supervisor would send them off in another direction with their mops. Whilst for the mall management, then, it was important to provide a disciplined workforce as a backdrop, the impact of the cleaners' presence was nevertheless strategically managed so as to produce the illusion of their invisibility. Most notably, in the gregarious, sociable atmosphere of the mall, where customers and shop assistants chatted openly, the cleaners worked alone and silently, the result of the employers' rules which stated that cleaners were not to chat to other cleaners or to form groups. Indeed, the cleaners' silence in the public spaces was quite noticeable in a country where the exchange of banter is a constant backdrop and an important aspect of social identity.

Although the cleaners were expected to be invisible to shoppers in the mall, they were nonetheless under constant surveillance. Thus, supervisors would make regular rounds, dressed in their smart blue uniforms, and any deviance could be immediately conveyed through walkie-talkies. The tracking of the cleaners' movements did, however, take on a new dimension when the company was being assessed for the international standards mark ISO 9001 2000, a dimension which led to some resentment among the women, a point I will build on later.

Additionally, just as space was strictly divided, so, too, was time. Hence, cleaners had to be "on time", clocking in before 8am registration every morning and clocking out twelve and a half hours later. The two 45-minute breaks were strictly timed and always taken on time.

This physical placing and erasing of the cleaners in these complex spaces was echoed in their social placing within a particular hierarchy. Thai society is extremely hierarchical, with respect and deference paid to older relatives and colleagues. In the workplace this attention to social place is reflected in the epithet "*phi*" (older sibling) being accorded to someone of higher status and "*nong*" (younger sibling) to someone who is considered lower in the social pecking order. For example, in a restaurant, it is customary to refer to waiting staff in a restaurant as *nong*, regardless of their age. These distinctions were deployed within the mall in ways that were made very clear to me on my first day of conducting research there. Thus, the assumption of many supervisors had been that I would eat my lunch with them inside the office, separate from the cleaners. My announcement that I wanted to eat with the cleaners in order to get to know them elicited raised eyebrows and some consternation: "You want to eat with the children (*dek dek*)?" one supervisor quizzed me.[6] Indeed, the moment I took my food over to the long trestle table where the cleaners were talking and laughing, I knew I had crossed an invisible boundary and allied myself with the "children"—the cleaners. At that instant, I realized that being simultaneously on both sides of the boundary would be impossible. Somehow, the choice I had made was a catalyst for my relationship with the cleaners, who understood that I was now "with them". Meanwhile, I felt that I was regarded from that point on with some suspicion by the supervisors and certainly was never rewarded with the frankness of information or genuine friendship I experienced with the cleaners. These distinctions were reflected in the daily ritual of registration, during which the cleaners would line up, with the supervisors and the chief of staff in front, and would be given information about various issues or sometimes chastised for poor work. In fact, starting the day with these regulated patterns was in many ways an embodiment of the cleaners' place *vis-à-vis* more senior members of staff and of their collective uniformity.

The irony in all this, however, was that the supervisors were only marginally better educated than the cleaners, often lived in the same slum community, and obviously did not receive a much higher salary. What distinguished the groups, then, were the types of work in which they engaged and the opportunities for public display—for being part of the mall's active, visible meanings—that the post of supervisor offered. Given that within the dominant, middle-class conception of moral worth and sophistication physical labor is connected with a lack of progress, coarseness, and backwardness, the supervisors' relatively languid work routines and ability to avoid manual labor clearly marked them as more "civilized" than those they supervised. Such attitudes are not unique to Thailand and, in fact, exist in many parts of Southeast Asia. Hence, Sen (1998) talks about the use of maids in Indonesia for the dirty, labor-intensive aspects of cooking that are conducted in a concealed "back kitchen" whilst the lady of the house prepares food for her guests in the spotless more public front kitchen, part of a social and spatial organizational structure designed to demonstrate her skills as a hostess. These attitudes are not necessarily restricted to the middle and upper classes, but may also be employed as a hierarchical distinguishing device amongst lower-class people. Thus, in the mall the supervisors' self-presentation and interactions were partly aimed at marking publicly their social distance from those who labored for a living. Indeed, although some of them would actually help out with the cleaning before the mall opened, the remainder of the day would be spent walking around their assigned floor. Their work seemed considerably less monotonous than that of the cleaners, and I often saw them chatting in the office or upstairs in the storeroom. In addition, they would spend a considerable amount of time putting on make-up and painting their long fingernails, which were clearly not designed for physical labor. By contrast, any hints of glamour among the cleaners were minimal, restricted to small touches such as a pair of earrings. The shapeless company shirts and blue trousers they wore further imposed a degree of functional uniformity on the cleaners.

Significantly, these hierarchical differences were quite gendered. Certainly, in some situations the collective body of cleaners was treated as a single, undifferentiated group. With both men and women identically dressed in the regulation blue trousers and shirt, with all being a similar height and stature and with all having short or tied-back hair, it was not always easy to distinguish men from women during morning assemblies. Yet, gendered distinctions did quite clearly emerge at particular points in at least two ways—concerning contact with dirt and workers' expectations of visibility and public display. First, it was the case that the female cleaners had to clean both the male and female toilets and when I asked why this was so, the cleaners were amused, telling me: "of course, men can't go into ladies' toilets". Certainly, it was entirely possible that the reasons for such practices were guided as much by the need to protect the modesty of the women using the facilities or by the Thai belief that men should not be close to women's genitals or menstrual blood—which are considered potentially damaging to their sexual and physical potency—as much as it may have been because of an unspoken, yet tacitly accepted, rule that makes direct contact with bodily waste the work of women. Whatever the reason (and I was never given one) the

existence of this gendered division of labor does illustrate how traditional mores—whether beliefs about female fluids or the continued
hold over the imagination exercised by longstanding notions of the
appropriate sexual division of labor—shaped work practices in this
edifice of modernity.

The second quite clear' gendered division of labor was marked by
the differences between the work of the cleaners and that of the
security guards. Although earning similar wages and usually only with
a slightly higher level of education, the guards—all male—were
nevertheless engaged in a job with more obvious status, a status
reflected in the smart, almost military-style, uniforms they wore and
their visibility as protectors of shoppers' personal safety. Although the
guards rarely had to exercise their authority, they played an important
role in maintaining the order of the mall, embodying authority as a
deterrent to obviously poor, scruffily-dressed people whose presence
in the mall might, in itself, be considered to be "matter out of place".

Assumptions About Rural Knowledge: Entering the System

Finally, aspects of the working environment reinforced perceptions of
the cleaners' place in a developmental hierarchy. This was expressed
through assumptions of the knowledge and behavioral traits the
cleaners might bring with them to the urban workplace. For instance,
when I asked why the cleaners had to undergo costly training upon
being hired, the cleaning company chief of staff at the mall—Daw—
explained: "Maybe they have done cleaning before, but there is more
to it here. It's not just a question of sweeping. They have to start
doing it very well, otherwise they can't enter the system". What is
significant about this comment is that the underlying thrust of Daw's
argument was that sophisticated, standardized, taught knowledge
provides the necessary transition from an unsystematic rural to a systematic urban approach to cleaning and, presumably, life in general.
These implicit expectations about behavior were supplemented by
the more concrete rules enshrined on the wall outside the office.
Thus, as noted above, the cleaners were not permitted to "chatter with
each other, joke around or shout whilst working". Furthermore, they
were not allowed to eat, drink, smoke nor to go into the shops in
the mall whilst working. Additionally, they were expected to pay
respect to their superiors and it was made clear that workers must
not "use or take company goods or products for [their] own use".
When I asked Daw why these rules were needed, she told me:

> We need rules because things are so different for people from the
> provinces (*tang jangwat*). They don't do things in the usual way. They
> don't have bosses. "Up-country" they are their own bosses. Here, there
> are supervisors . . . It's not our house; they employ us to come and

work, so we need rules. Many things are forbidden, such as drinking alcohol at work. Stealing is forbidden. Some people who come from up-country don't know—they see something nice and they take it. It's forbidden to eat food upstairs because it doesn't look good in front of the customers.

Daw's distinction between normalcy and abnormalcy—"They don't do things in the usual way"—clearly marked "up-country" ways as abnormal/unusual and implicitly suggested that training and adopting the standards of city life and work would bring rural migrants into the modern world.

Claiming Small Freedoms: The Cleaners' Responses
In this final section I explore how the cleaners sought possibilities for expression and flexibility within the controlled spaces and relationships I have described above. I argue that these subtle processes of reclaiming themselves in the face of uniformity and structure were forms of "everyday politics", a reaction to the immediacy of restrictions, as well as a commentary on the wider politics of modernization.

Pragmatism and Working-Class Pride
What I learned from spending time with the cleaners was that they appreciated certain aspects of the cleaning work, compared with previous situations in which they had found themselves. One major advantage was that the work was inside in the air-conditioned mall, rather than in the hot glare of the sun that would make their skin "black" which, in Thai representations of ethnicity, would mark them as having low-class status. Given that most of the women had experienced farm labor, which would mean spending all day in the sun, whilst others had worked on construction sites in Bangkok and other places—all locations of outdoor work requiring heavy physical labor—they saw being employed in the air-conditioned mall as a definite step up. Another benefit they perceived was the flexibility of the work: as noted above, not being contracted meant that they could go back to their villages at harvest time or to visit their families for extended periods, knowing that they would be able to work on their return to Bangkok, since the company showed good workers loyalty and needed to compensate for a relatively high turn-over of staff.[7] I also discerned what might be termed "working-class consciousness", particularly from the Isan women, who often stressed to me the centrality of being strong and capable of hard work as part of their expression of an Isan identity. The ability to "withstand hardship" (*oton wai*) was a recurring theme in my conversations with the

women, and comparisons would often be made with the "weak" and "lazy" Bangkok people, who did not know the meaning of hard work. Indeed, there was an underlying pride in being able to do a good, honest job and in being self-sufficient against a background of hardship.

These attitudes in the context of the low-status cleaning work were succinctly expressed in the life story of Tuk, an Isan cleaner in her late twenties who was a single mother and supported a child and her elderly mother, both of whom were living in rural Isan. She indicated that some of the women she had worked with were ashamed to be cleaning toilets for a living. These women's fickle attitudes became a foil for Tuk's own attitudes to work, and for the strength of character she saw as integral to her own survival and success. She told me:

> the people from Bangkok think we're from up-country and we don't have any knowledge, and they like to look down on us people who do low-level work. But I don't care. Those who care can't put up with (*oton*) the work, and they leave, but I don't care . . . I don't like it when other people look down on themselves and say: "why do I have to do this kind of work?" If I hear someone talking like that, I ask: "Do you think you're so high class that you can't do this work?" Some people think it's too hard, but I think if you want to do it, you have to be able to put up with it.

Significantly, there are parallels here with Constable's (1997) work on Filipina maids, whose professional attitudes were empowering in their capacity to bring job satisfaction and—perhaps—to strip their employers of any justification for critique. Yet there were other underlying motives for the women's hard work and desire not to jeopardize their livelihoods. In my conversations with the women it was clear that they took pride in their work because of the choices it enabled them to make in other areas of their lives. Thus, unlike in some Asian countries, rural Thai women play a central role in production. Not only do they actively participate in farming activities, but they often also run businesses. It is women who control family finances and who are expected to contribute financially in the reproduction of their families. Indeed, as I have noted elsewhere (Brody 2003), being able to provide for parents and children is an important gauge of successful womanhood and personhood. Many of the women whom I interviewed had family and land back in their home villages, and the money they saved was used for their children's education and for farming. Hence, their work was enabling them to ensure a better future for children who would support them when they themselves were unable to work. It was also a means to actively use and benefit from their land, so retaining a sense of ownership over the soil of a country that is increasingly being defined from its urban center. In this complex relationship of the cleaners' work and their vision of self-sufficiency, a sense of freedom came from their belief that this was a step to something better.

Commentaries on Modernization: Creating Spaces for Fun

As I have noted above, the work itself was not necessarily a cause for discontent. What the women complained about constantly and resented were the rules and strict boundaries constraining their freedom. They did not appreciate being told what to do and how to do it. Their frustration often found expression in nostalgia for a rural lifestyle that was equally hard and certainly not secure, but whose benefit was that it was self-directed and therefore bestowed a sense of freedom (*khwam issara*), however illusory. In this vein, Hirai (1998:29), in a study of Northern Thai factory workers, has made much of the differences between "village" and "city" work, noting that the origin of the word *tamngan* (work) is *ngan* (festival/communal activity), a concept that is concerned with the fostering of social relationships rather than economic activities. This traditionally rural concept of work is embodied in harvesting, a time when rural villagers provide mutual support for the gathering of each other's harvests. Although it is hard work, it is also "*sanuk*" (fun), a very important principle in all Thai activities. I experienced this myself during harvesting in Nan province with Gulab, one of the cleaners in the mall, and appreciated the social aspects of the work, which was hard but which could also be an opportunity for chatting or flirting. Most of Gulab's family had returned from Bangkok for her wedding, so the rice fields were also a point for reunion. There were breaks for laughter and songs, and in the midday food was eaten communally under shady trees until it was cool enough to work again.

This quotidian rhythm contrasted with work routines in the mall. Indeed, Gulab was particularly vocal about the restrictions on her time and movements there, and she yearned for the life at home where she could work and eat when she wanted. She told me:

> "[At home] whatever you want to do, you can. You have freedom. I was really *sabai* [felt good] at home. I went to the fields [to work] everyday . . . When I came home I relaxed. Some days we had a rest or went for *tiaow* [a trip] nearby by motorbike. I was free, but I didn't have any money". She was thus torn between the need or desire to make some money through the links that she had with the cleaning company, and the desire for quality of life she perceived at home. This dilemma was heightened when she had a child, whom she desperately wanted to stay with and care for, but knew she could not raise without the money that came from her work (for parallels with South Africa, see Bezuidenhout and Fakier in this collection).

There were, however, ways to claim small freedoms within the restrictions of the cleaners' routines, and these grasped opportunities were significant within the overall context of work. One way to quietly subvert the structured, disciplined work regimes was to appropriate spaces within the mall for *khwam sanuk* (fun), enabling the cleaners to impose, to an extent, their own interpretations of work. Thus, although the cleaners were not supposed to chat together during working hours, the toilet areas and some of the larger storage cupboards, situated away from the main shopping areas of the mall and from the eyes of the supervisors, often provided an arena where gossip could be exchanged. For instance, Wai, a 43 year-old cleaner from Isan, loved relating amusing anecdotes. She was rarely alone at her station because the employees at the bank next door could not resist talking to her. The storage cupboard by the restrooms on one floor, then, provided a haven for women's talk, talk in which certain supervisors sometimes also participated. Flirtations were also played out in these spaces. Thus, Gulab met her new husband, a security guard at the mall, because he was stationed by the restrooms for which she was responsible. During their courtship they would joke together as he passed by, and after their marriage he would regularly visit her at her post. When surveillance of the cleaners was increased because of the ISO 9001 2000 assessment, some of these small freedoms were undermined, creating an atmosphere of discontent and resentment.

The fact that the slum community where many of the women lived was right next door to the mall also offered opportunities for reclassifying spaces.[8] In a way similar to how the Filipina maids about whom Constable (1997:202) writes were able to domesticate certain public spaces of Hong Kong during their free time, exhibiting public displays of "loud, uninhibited behavior" which are frowned upon in Chinese society and which she takes to be an expression of quiet resistance and freedom from the restrictions of the workplace, many of the people from the slum (both cleaners and non-cleaners) would sit outside the mall in the cool of the trees, on the comfortable marble steps, watching their children play. Others would profit from the presence of tourist buses and middle-class shoppers to sell items such as cheap bags or toys.

Finally, the cleaners would often go to the slum community itself for breaks, sometimes ordering *Somtam*, a typically Isan dish made of raw papaya. This struck me as one means for the cleaners to feel re-humanized in a space where they were known as people with lives, relationships, and personalities, rather than as part of a faceless workforce. By walking next door, they entered a world within whose meanings they actively participated and where they were not surreptitiously monitored for deviance.

Such expressions of sociality and alternative relationships carved out of an otherwise humorless regime were, I suggest, more than a way for the women to survive the tedium of their long days. They also marked a degree of contempt for the repressive values of those who employed them and which were imposed upon them by virtue of their place of relative disempowerment in an unequal system of distribution. Certainly, the women were reliant on the work and on those who employed them, but they yearned for their own children to eventually escape these relations of subservience. Indeed, by embracing values that have been written off as "backward" and no longer valid in official representations of modern Thailand, they were demonstrating that such values nevertheless do have continuing currency and significance for them. In fact, paradoxically, it was their work in the city that was enabling them to maintain ownership of rural spaces and livelihoods, and to defend local culture which, ultimately, was more meaningful to them than was the alienating environment of the city that is being held up by government planners and others as the pinnacle of modern achievement. In this way, then, they quietly asserted an alternative vision of the future for their country, one rooted in local values of mutual trust, community, and working the land.

Conclusion

In this paper I have considered the centrality of dirt, cleanliness, and discipline in creating an image of modernity in a Bangkok shopping mall, and in marking its diametric opposition to the "less civilized" space of a rural village. In the mall, the presence of cleaners reinforces the message that consumers have entered a space that is commensurate with their own actual or desired level of progress and sophistication, although the presence of such cleaners, ready to banish dirt and maintain order, is at times more symbolic than anything. At the same time, I have suggested that the erasure of the cleaners as people, through strict codes and constant surveillance, is a corollary of the public/private divisions exhibited in many aspects of middle-class Asian life, such as Sen's (1998) observations concerning Indonesian hostesses' reliance upon maids to help with dinner parties—observations which themselves match Goffman's (1959) famous analysis of how social actors seek to present an idealized version of "reality" to the public whilst concealing much of the work and/or workers necessary to make the representation work. Thus, a central aspect to maintaining the image of the mall as a space of modernity for its customers is to ensure that the cleaners who remove its dirt work behind the scenes. The underlying message here, then, is that the bodies of poor, "less civilized" rural people must be surveilled, made invisible, and—if necessary—corrected, so that they are not perceived to be a threat to the image of the mall as a modern environment.

I have also shown that the cleaners have their own perspective on the work they do, and are sustained by the opportunity the work offers to assist their families and to be self-sufficient, as well as valuing their own capacity for hard work. The cleaners did not, however, readily accept certain aspects of their working environment, which they recognized as patronizing and restrictive of their freedom and self-expression. Indeed, these women constantly sought small freedoms as a way to reclaim themselves against efforts to diminish their presence and their individuality. Certainly, these tiny acts of rebellion cannot be compared with, say, large-scale union activity, but they are significant when considered against the complex web of meanings described in this paper. Thus, following Scott (1985), they can be read as part of a "hidden transcript" which constitutes the cleaners' "ordinary weapons" of resistance and provides a means for them to assert themselves in ways that confer a sense of dignity, autonomy, and ownership. Without doubt, most of the cleaners would not wish to confront their employers directly, for fear of losing their jobs—a fear particularly felt by older women who cannot move easily between jobs and so, unlike younger cleaners, cannot simply leave if they became frustrated or bored. However, for those who wished to retain their jobs, seeking ways to slip between the rules provides a means to make their daily lives more bearable. This is significant, because the discourse of modernization within which such cleaners are caught assumes that the cleaners will be affected by their contact with urban modernity, whereas the spaces of urban modernity will remain unchanged—that is to say, that "civilization" and "modernity" will affect the cleaners but the cleaners will not affect how "civilization" and "modernity" manifest themselves.

Hence, as I have argued in this paper, behind the deployment of these everyday politics was a deeper political undercurrent, whether consciously intended or not. Thus, the use of the women's earnings to support rural livelihoods, coupled with their unwillingness to comply with the structured patterns dictated by the capitalist ethos of the mall management, set up an implicit challenge to the idealized image of a modern, largely urbanized Thailand driven by transnational businesses. Certainly, the women accepted the need to work for a company that depersonalized them and that depended on their labor yet that kept them in their place. However, at the same time, they aspired to autonomy and ownership, planning entrepreneurial activities such as selling their own produce or buying and reselling goods, and putting money into homes and land in their rural communities. In these ways, the cleaners' attitudes disputed notions of cultural uniformity that underpin the project of constructing a modern Thai nation. Moreover, their lack of respect for the rules was an indication that they knew those making the rules could only do so by virtue of an unfair, unequal system determined by power rather than merit. Many of them saw their children's education as a passport to jobs as managers and bureaucrats—making, rather than tolerating, the rules. They also provided their own interpretation and appreciation of the place and role of rural areas, lifestyles and people, and—above all—of the Thai laboring classes. In these ways, then, the cleaners offered insights into a more fractured Thailand than the continued project of constructing a monolithic modern "Thai-ness" would suggest.

Endnotes

[1] In 1997 recorded migrants in Bangkok were 196,838 out of a population of 7,238,953, though the likelihood is that the real figure was double this number, due to the number of unrecorded migrants (Thai National Statistical Office 1997).

[2] The theme of repaying a debt to parents in the form of being "dutiful" is a common one in Thailand and has been dealt with extensively in the available literature (see Phongpaichit 1982).

[3] For discussions on urbanization in Thailand, particularly in Bangkok, see Askew (1994, 1996) and Korff (1989).

[4] See Wilson (2004:29–67) for a history of the Thai shopping mall.

[5] The International Organization for Standardization (ISO) is a global federation of national standards bodies established in 1947. Its mission is to promote the development of global standards so as to facilitate the international exchange of goods and services, and to develop cooperation in the spheres of intellectual, scientific, technological, and economic activity. ISO 9000 is a set of standards for quality management systems. The suffix 2000 refers to revisions in the standards which took effect in that year.

[6] The term *dek dek* is evidently not limited to the working environment I observed. Hirai (1998) also noticed that this distinction was made between ordinary workers and more senior workers in the factory where he conducted his fieldwork.

[7] The high price of living in Bangkok drew some of the women I knew back to their home villages during my research period. Others left for a few months to assist with the harvest or to give birth, whilst a few younger women sought work elsewhere—in factories or in other service industries, where their youth was seen as an asset.

[8] I do not have space here to talk about the centrality of slum communities for migrant workers in Bangkok. The communities, though characterized by poor quality housing and poor sanitation, nonetheless offer several advantages for those living there. Thus, when they are centrally located, they enable workers to cut down both the cost and the time involved in commuting from other areas. They also provide a ready-made support group of people in similar situations, often from the same rural areas, whilst offering a cheaper parallel economy for migrants, enabling them to buy food and other commodities and services at lower costs than outside the community (see Rabibhadana (1975) for an excellent ethnography of slum culture in Bangkok).

References

Askew M (1994) *Interpreting Bangkok: The Urban Question in Thai Studies*. Bangkok: Chulalongkorn University

Askew M (1996) The rise of moradok and the decline of the yarn: Heritage and cultural construction in urban Thailand. *Sojourn: Journal of Social Issues in Southeast Asia* 11(2):183–210

Barmé S (1993) *Luang Wichit Watanagan and the Creation of a Thai Identity.* Singapore: Institute of Southeast Asian Studies

Boonchalaksi W and Guest P (1994) *Prostitution in Thailand.* Bangkok: Institute for Population and Social Research, Mahidol University

Boonmathya R (1997) "Contested concepts of development in rural northeastern Thailand." Unpublished PhD dissertation, Department of Anthropology, University of Washington

Bourdieu P (1977) *Outline of a Theory of Practice.* Cambridge: Cambridge University Press

Brody A (ed) (2000) *Uniting Voices: Asian Women Workers' Search for Recognition in the Global Marketplace.* Bangkok: Committee for Asian Women

Brody A (2003) "Agents of change: Struggles and successes of Thai women migrants in Bangkok." Unpublished PhD dissertation, School of Oriental and African Studies, University of London

Chin C (1998) *In Service and Servitude: Foreign Female Domestic Workers and the Malaysian "Modernity" Project.* New York: Columbia University Press

Chouwilai J (2000) Life after redundancy: The effects of the termination of Thai workers. In A Brody (ed) *Uniting Voices: Asian Women Workers' Search for Recognition in the Global Marketplace* (pp 85–98). Bangkok: Committee for Asian Women

Constable N (1997) *Maid to Order in Hong Kong: Stories of Filipina Workers.* Ithaca, NY: Cornell University Press

Douglas M (1966) *Purity and Danger: An Analysis of the Concepts of Pollution and Taboo.* London: Routledge and Kegan Paul

Escobar A (1995) *Encountering Development: The Making and Unmaking of the Third World.* Princeton: Princeton University Press

Evans G (1999) *Laos: Culture and Society.* Chiang Mai: Silkworm

Foucault M (1975) *Discipline and Punish: The Birth of the Prison.* London: Penguin

Fuller T, Kamnuansilpa P, Lightfoot P and Rathanamongkolmas S (1983) *Migration and Development in Modern Thailand.* Bangkok: Chulalongkorn University

Fuller P and Lightfoot P (1984) Circular migration in Northeastern Thailand. In H Brummelhuis and J Kemp (eds) *Strategies and Structures in Thai Society* (pp 81–108). Amsterdam: University of Amsterdam

Goffman E (1959) *The Presentation of Self in Everyday Life.* Garden City, New York: Doubleday

Harvey D (1989) *The Urban Experience.* Oxford: Blackwell

Hirai K (1998) "Women, family and factory work in Northern Thailand: An anthropological study of a Japanese factory and the villages of its workers." Unpublished PhD dissertation, Department of Anthropology, London School of Economics

Hirsch P (1990) *Development Dilemmas in Rural Thailand.* New York: Oxford University Press

Hoy S (1995) *Chasing Dirt: The American Pursuit of Cleanliness.* New York: Oxford University Press

Kemp J (ed) (1989) *Peasants and Cities, Cities and Peasants: Rethinking Southeast Asian Models.* Overveen: ACASEA

Korff R (1989) *Bangkok and Modernity.* Bangkok: Chulalongkorn University

Mills M B (1993) "'We are not like our mothers': Migrants, modernity and identity in northeastern Thailand." Unpublished PhD dissertation, Department of Anthropology, University of California, Berkeley

Mills M B (1997) Contesting the margins of modernity: Women, migration, and consumption in Thailand. *American Ethnologist* 24(1):37–61

Mills M B (1999) *Thai Women in the Global Labour Force: Consuming Desires, Contested Selves.* New Brunswick: Rutgers University Press

Ngaosyvathn M and Ngaosyvathn P (1994) *Kith and Kin Politics: The Relationship Between Laos and Thailand*. Manila: Contemporary Asian Publishers

Odzer C (1994) *Patpong Sisters: An American Woman's View of the Bangkok Sex World*. New York: Arcade Publishing

Office of the Prime Minister (1996) *Eighth National Economic and Social Development Plan (1997–2001)*. Bangkok: Thai Royal Government

Office of the Prime Minister (2001) *Ninth National Economic and Social Development Plan (2002–2006)*. Bangkok: Thai Royal Government

Phongpaichit P (1982) *From Peasant Girls to Bangkok Masseuses*. Geneva: International Labour Organisation

Phongpaichit P and Baker C (1995) *Thailand: Economy and Politics*. New York: Oxford University Press

Phongpaichit P and Baker C (1998) *Thailand's Boom and Bust*. Chiang Mai: Silkworm Books

Pigg S L (1992) Inventing social categories through place: Social representations and development in Nepal. *Comparative Studies in Society and History* 34(3):491–513

Rabibhadana A (1975) "Bangkok slum: Aspects of social organization." Unpublished PhD dissertation, Department of Anthropology, Cornell University

Rajadhon P A (1972) *Watthanatham (Culture)*. Bangkok: Samnakpim Banakarn

Reynolds C (ed) (1991) *National Identity and its Defenders: Thailand 1939–89*. Chiang Mai: Silkworm Press

Scott J (1985) *Weapons of the Weak: Everyday Forms of Peasant Resistance*. New Haven: Yale University Press

Sen K (1998) Indonesian women at work: Reframing the subject. In K Sen and M Stivens (eds) *Gender and Power in Affluent Asia* (pp 35–62). London and New York: Routledge

Thai National Statistical Office (1997) *Thailand Statistical Handbook*. Bangkok: Office of the Prime Minister

The Nation (Thailand) (2000) Row deepens over star's TV insult. 3 April

Thitsa K (1980) *Providence and Prostitution: Image and Reality for Women in Buddhist Thailand*. London: CHANGE International Reports

Thorbek S (1987) *Voices from the City: Women of Bangkok*. London: Zed Books

Turton A (1984) Limits of ideological domination and the formation of social consciousness. In A Turton and S Tanabe (eds) *Historical and Peasant Consciousness in South East Asia* (pp 19–73). Osaka: National Museum of Ethnology.

Turton A (ed) (2000) *Civility and Savagery: Social Identity in Thai States*. Richmond, Surrey: Curzon

Ungpakorn J (1999) *Thailand: Class Struggles in an Era of Economic Crisis*. Hong Kong: Asian Monitor Resources Center

Wilson A (2004) *The Intimate Economies of Bangkok: Tomboys, Tycoons and Avon Ladies in the Global City*. Berkeley: University of California Press

Chapter 8

Cleaning Up After Globalization: An Ergonomic Analysis of Work Activity of Hotel Cleaners

Ana María Seifert and Karen Messing

Introduction

In the hospitality industry, neoliberalism and increasing industrial concentration have recently become ever more visible. As competition in the industry has deepened, hotel chains have felt pressure to become large enough both to satisfy the travel needs of their increasingly wide-ranging clients and to develop the economies of scale that will improve their bottom lines and so satisfy their shareholders. This growth requires cash from global capital markets, so that the majority of hotel chains now issue stock publicly (Bernhardt, Dresser and Hatton 2003). However, publicly owned corporations are required to report to stockholders at frequent intervals in order to demonstrate good returns to investors (Cline nd). This leads to significantly increased pressure for short-term performance rather than long-term sustainability, resulting in organizational restructuring to cut costs and to increase revenue flows (Bernhardt, Dresser and Hatton 2003). To improve their immediate bottom line, then, hotels and hotel chains have sought to introduce more "flexibility" into their operations in the form of part-time, casual, and seasonal work, a strategy which has significantly diminished the attraction of investing in the long-term health or job satisfaction of workers. At the same time, hotel managers have outsourced functions such as laundry and catering services. Moreover, they have increased efforts to woo and render faithful a fickle, sophisticated global clientele by offering more amenities in their guest rooms (coffee pots, hair dryers, irons and ironing boards, bathrobes, extra sheets and pillows, and even printers and fax machines; Bernhardt, Dresser and Hatton 2003; Seifert 2001). As the hospitality industry has responded to globalization and increased competition through new marketing initiatives, employment practices, and restructuring decisions, the work of cleaners has both dramatically intensified and their employment has become significantly less stable (Scherzer, Rugulies and Krause 2005; Seifert 2001).

The research presented here will explore how hotel cleaners' work is being reshaped by these new work regimes. We report on two ergonomic case studies of cleaning in hotels in Montréal, Canada, initiated to identify determinants of the difficulties in cleaners' work in order to suggest improvements to their working conditions. These studies included observations, interviews, and documentary research. Based on this research, we first present a general description of Montréal hotel workers and their work, and we identify a number of factors that are intensifying this work and leading to conflicts among cleaners within the workplace. We also describe strategies used by workers to deal with these new constraints. Finally, we explain how the union has used the results of these studies in an effort to improve working conditions. In conducting the research we draw upon a tradition of ergonomic analysis as developed in the French-speaking world which has been used to reveal relatively invisible constraints and requirements in the work environment, as well as the strategies used by workers to deal with them (Guérin et al 1997; Teiger and Bernier 1982). We have previously been involved in efforts to apply this type of analysis to women's work in the service sector (for example, Seifert, Messing and Dumais 1997). For us, cleaning is a particularly interesting type of work for such analysis for several reasons: it is subject to a fairly strict sexual division of labour (Gaucher 1981; Messing, Haëntjens and Doniol-Show 1993; Messing, Chatigny and Courville 1998); it is considered to be a traditional task for women; and even though cleaning is associated with significant risks for musculoskeletal problems (Milburn and Barret 1999; Torgen, Nygard and Kilbom 1995) and with health problems derived from exposure to cleaning products (Karjalainen et al 2002; Rosenman et al 2003), many employers and even some union activists believe that cleaning rooms is a natural and undemanding activity, done without problems by many women.

Despite its importance for human health promotion (cleanliness is, after all, an important aspect of a healthy environment), cleaning has often been seen as peripheral to the "real" goals of production or client services and therefore cleaners have often been excluded from ergonomic studies and interventions in the institutions, factories, and retail establishments where they work. This study, then, is intended to help fill that gap.

An Overview of Hotel Cleaning Work

The economic dimension of globalization is manifested in the hospitality industry in Canada in at least two significant ways: first, as in many parts of the world, mergers and acquisitions have brought about a concentration of ownership in the hands of a small number of highly competitive transnational corporations; second, international migration has brought many workers from the global South to metropolitan centres to fill jobs that are too poorly paid to attract many local workers. These developments are having consequences both for the nature of cleaners' work and for the emerging work relationships amongst cleaners. As a result, unions are scrambling to develop strategies that will allow them to respond to cleaners' worsening working conditions and to encourage solidarity amongst ethnically diverse workers despite organizational practices that have often pitted one ethnic or cultural group against another.

Cleaner Demographics and the Low Status of Cleaning

In Montréal, the politics of work, class, and race/ethnicity have intersected in important ways as globalization and neoliberalism have impacted the hotel industry through the introduction of new labour regimes and a growing reliance upon new groups of (immigrant) workers. Today hotel cleaners are primarily immigrant women, with a majority being either black or Latin American. Such demographics in the industry have become increasingly common as workers from developing countries have migrated to the US, to Canada, and to Western Europe. Hence, in studies of hotel cleaners in San Francisco and Las Vegas, Lee and Krause (2002), and Scherzer, Rugulies and Krause (2005) found that almost all cleaners were immigrant women, as did Muqa et al (1996) in Paris, France, where a majority of hotel cleaners were immigrants. According to Glenn (2001:73), this growth in reliance upon immigrant workers has been "concomitant with globalization" as economic restructuring has forced "women from the periphery to migrate to metropolitan centers to fill demands for both private and public reproductive services". The result, she suggests, has been that in countries such as the United States, black and Latina women have been "disproportionately employed as service workers to carry out low-level public reproductive labor".

Given that cleaning is generally acknowledged to be low in status, cleaners are poorly paid, their workload is generally not recognized, and their work is usually seen as peripheral to the main activity of the employer (Messing 1998). Cleaners often complain of a lack of respect shown them, a lack which seems to be the result of a number of factors, one of which appears to be cultural associations between cleaners and the dirt they must remove (Glenn 1992; Brody this collection). Such disdainful behaviour on the part of other staff and of hotel guests is not just unpleasant, but it also has considerable impacts upon cleaners' health—it can, for example, undermine their mental health and lead to anxiety and low self-confidence, both of which are linked to psychological damage (Cortina et al 2001; Dejours 1993; Leary and Kowalski 1995:137). At the same time, cleaners' low status can result directly or indirectly in damage to their physical health, the result, usually, of their not being consulted by administrators with regard to the design of the spaces and furnishings they clean (Messing, Chatigny and Courville 1998). Hence, even when equipment is ordered specifically for their use, cleaners are usually not involved in its choice and do not pre-test it. For instance, in the research recounted here, a supply cart was too heavy and the push bar was too high for most of the hotel cleaners we observed working. Such poor ergonomic design can have considerable implications for physical injuries to the body, over both the short and long term, yet these and other physical challenges of cleaning are generally underestimated.

Perceiving oneself as part of a low-status group can also lead to tensions and aggression among those who experience contemptuous treatment (Wilkinson 1999). Indeed, in the Montréal hotels we studied, the women cleaners were perceived by managers as a jealous and conflict-ridden group. Such behaviour has been aggravated by the fact that cleaners' work is usually done under significant time pressure, a time pressure which has been exacerbated (as we will show here) by the amplified physical, mental, and emotional workload linked to procedures introduced recently by employers wishing to show increased short-term profits.

Race and Class Solidarity and Conflicts

Whereas historically hotel cleaning in Montréal was dominated by white, Canadian-born workers who enjoyed relatively long-term employment relationships with hotels, as globalization has impacted migration and working patterns, the population of hotel cleaners in Montréal has become a complex mosaic composed of working-class white women, working-class immigrant women, and educated immigrant women, many of whom now work on a contingent (ie part-time, temporary, and/or sub-contracted) basis. This composition involves cleaners in dynamics of race and class solidarity, as well as conflict. For example, we have observed that racial/ethnic alliances are used both to regulate workload and to defend group members against perceived injustice. Equally, class consciousness also contributes to solidarity against perceived injustice to a cleaner and feeds into daily and long-term struggles for better working conditions. However, our analyses of the industry show that there are also multidimensional cleavages between "Canadians" and "immigrants", between "regular cleaners" and "students", and between more-senior and less-senior cleaners. These cleavages cause rivalries and conflicts that can diminish solidarity and greatly shift the power balance in favour of administrators.

Certainly, such cleavages of race and class are not unique to hotel cleaning, and Glenn (1992:34) and Scherzer (2003) have shown how they shape conflicts between registered nurses (usually white and middle class) and nurses' aides (more often minorities) in the United States over such matters as work assignments. In the Montréal hotels we have studied, though, conflicts amongst nationalities are quite common and can be partly attributed to class conflict between the more-educated, Canadian-born students and immigrants, and the other workers. Thus, white Canadian-born students who take up cleaning as a part-time or summer job whilst they are studying tend to speak readily with administrative personnel, an attitude that is perceived by many non-student cleaners as a betrayal of cleaners' class interests. Likewise, educated immigrants sometimes try to gain some recognition by bringing up their previous educational experiences and by expressing opinions in an articulate fashion, behaviour which is experienced as contemptuous by many of those with less education, both working-class, white, Canadian-born cleaners and immigrants.

Class, race/ethnicity, geographic origin, and seniority, however, do not act alone. Rather, it is the flexibilization of work, the economic insecurity, and the exacerbation of differences in employment status (such as between more- and less-senior cleaners) caused by it that has encouraged rivalry and suspicion among cleaners. For instance, in our analysis of the industry we have seen that less senior cleaners are particularly worried about their number of work hours because their hours are quite variable and they therefore try especially hard to be perceived by housekeeping supervisors as "good cleaners". However, in doing so they then run the risk of being perceived by more senior cleaners as subservient and lacking class solidarity.

The Intensification of Hotel Cleaning Work
The intensification of hotel cleaning has been manifested in work content in three principal ways: a modified working environment due to increased efforts to secure market segment; changed employment contracts so as to bring about greater flexibility; and outsourcing of work previously done in-house. Each of these is having important impacts upon the cleaners.

Marketing Efforts
As competition for guests—particularly those on business trips who are perceived to have more money to spend—has deepened due to the pressures placed on hotels by stockholders and potential stock purchasers for ever greater profit margins, hotels have provided increasingly lavish amenities to entice travellers to stay with them

rather than with their rivals. Such efforts at marketing have shown up in hotel rooms in the form of larger numbers of gadgets to check, clean, replace, and tidy up, including coffeemakers, trays of cosmetics or food products, and ironing boards. These amenities are often featured at hotels' websites and are promoted heavily in their marketing campaigns. For example, in a study of eight large hotels in the United States, administrators told researchers they had added amenities specifically to compete for clients (Bernhardt, Dresser and Hatton 2003). Other marketing strategies are also having an impact on cleaning—the provision of bigger beds, heavier mattresses, and a third bed sheet (see study results below) have also dramatically increased the physical effort involved in cleaning rooms.

Employment Flexibility and Non-standard Work

Employment flexibility is not created by globalization, but globalization favours its development by increasing competitive pressures. This is having important consequences for workers. Thus, flexible employment practices have decreased the numbers of full-time workers and varied the number of working hours per week in the hotel industry in both Canada and the United States (Bernhardt, Dresser and Hatton 2003). In the hotels where this study was conducted, these practices have meant that workers' room assignments often change from day to day and week to week, so that cleaners are less able to regulate their workloads by postponing operations from one day to the next or to anticipate heavy workloads by doing some operations the day before. Although the effects of these employment policies are most evident for casual and part-time workers who do not have an assigned group of rooms, they also affect more-senior workers who have an assigned group of rooms because all cleaners have seen their workloads increase and have had to integrate into their usual workday those jobs previously done out of season. By requiring employees, especially less-senior employees, to work on irregular schedules, only occasionally, or to be constantly on call, such employment practices can not only have negative health effects (Quinlan, Mayhew and Bohle 2001) but also greatly increase the difficulties related to balancing work and family responsibilities. Hence, Prévost and Messing (2001), for example, have shown that telephone operators working on irregular and unpredictable schedules are constantly required to rearrange child care, day care for elderly parents or sick relatives, and children's activities. Furthermore, the need to be constantly "on call" actually exacerbates cleaners' poverty levels—workers who are "on call" cannot use their non-work time for alternate employment yet they do not get paid if they are not actually called into work at the hotel.

Significantly, though, whereas employment flexibilization is often presented as a way for employers to use labour more efficiently, the lack of predictability in cleaners' work schedules also appears to have hidden detriments for the hotels themselves, as cleaners' inability to plan ahead often means that they actually take longer to complete certain tasks. This is a finding which replicates that of other industries such as healthcare, where studies show that nurses who are scheduled on an unpredictable and/or occasional basis are generally rendered much less efficient in caring for patients, since they are obliged to gather information on patients *de novo* every time they enter a new ward (see Seifert and Messing 2004).

Outsourcing

The outsourcing of work previously conducted in-house can have significant impacts upon cleaners' work, even if their own jobs are not outsourced. Thus, as Quinlan, Mayhew and Bohle (2001) have argued, outsourcing often leads to considerable changes in the ways tasks are carried out because the contracts with outside providers may not make explicit all of the necessary—yet often invisible and informal—aspects of the tasks as they were conducted in-house (cf Messing, Chatigny and Courville 1998). At the same time, however, the hotel typically assumes that all parts of the newly outsourced tasks have, in fact, been addressed as before. The result is that many problems start to fall between the cracks.

Methods for Work Analysis at Two Montréal Hotels

In this section we turn to an analysis of how the kinds of work intensification associated with neoliberalism are impacting cleaners in two Montréal hotels. In researching the impact of job transformation upon these cleaners, we used methods derived from those developed at the Conservatoire national des arts et métiers in France (Guérin et al 1997). These methods are based on close observation of cleaners' work and interviews with them about their work, as a means of examining workplace dynamics and gaining a deeper understanding of the factors shaping the redesign of work. Our method has several stages: (i) defining and understanding workplace needs and forming a study committee; (ii) observing cleaners work and asking them questions about what they are doing; (iii) systematic analysis of these data; (iv) giving feedback to the cleaners; and (v) the production of a final report (Messing et al 2005).

Study Setting

Since 1993, our research centre has been engaged in a formal partnership with Québec's three major trade unions to study health and safety issues related to women's work (Messing and Seifert 2001). The unions use the results of our studies to improve working conditions and to help them in their political work. The studies presented here were supported by this collaborative effort. They were carried out in two large hotels, both of which are owned by multinational corporations. The first study was initiated in 1998 in a 400-room hotel (Hotel East) at the request of a Workers' Compensation Board rehabilitation officer. The second was implemented under contract to the hotel administration but was begun in response to pressures brought to bear by the local union after completion of the first study. It was conducted in 2000 at an even larger, 600-room hotel (Hotel South).[1] Some funds for the research were negotiated by the union and came from the employer. Employers were aware of our collaboration with the unions and accepted union participation in the definition and conduct of the study. They also discussed solutions with the unions and allowed public disclosure of the final report, as long as the hotel's name was not mentioned. In both hotels the project started with a meeting with an employer representative. We asked for their views on the need for a study and whether they felt there were any problems with regard to working conditions. At our request, joint labour–management committees were set up and were composed of both employer and union representatives, as well as the heads of the housekeeping and maintenance services. The role of these committees was to comment on the study objectives, its conduct, and, eventually, the solutions proposed.

The employers at both hotels first requested that we train room cleaners in proper work methods as a way to prevent musculoskeletal problems. Whilst considering this request, we first analysed the needs and perceptions of the various actors by interviewing them and through engaging in preliminary observations of work activity. We also conducted one-hour individual interviews with the director of human resources and with the head of housekeeping. These interviews allowed us to understand managers' perceptions of problems in housekeeping, as well as the criteria they used when assigning rooms to housekeepers. Twenty-minute individual interviews with room cleaners allowed them to identify difficult tasks and causes of pain, as well as the working conditions they wanted to change. We interviewed all the room cleaners who were at the workplace during two consecutive days: 20 room cleaners of the 46 in Hotel East and 32 of the 80 in Hotel South. At this stage, we also directly observed ten room cleaners working for a total period of 56 hours. We obtained information concerning national origin either from the cleaners themselves or from their friends, whilst information on sex, age, and seniority was drawn from employers' lists. Upon completion of this preliminary data gathering, we met with the study committee and outlined the many demanding tasks identified through this process. The study committee then agreed to change the orientation of the study from "training cleaners in good work methods" to "identifying determinants of some problems in housekeeping".

Through a constant back-and-forth process of observation, inter-pretation, and feedback between ourselves and the cleaners we were able to gain a deeper understanding of the dynamics of their work. We then observed eight room cleaners during an additional 40 hours in Hotel East and ten room cleaners during an additional 50 hours in Hotel South. We also analysed 60 inspection reports on room cleaning in Hotel East. These observations and interviews were conducted between September 2000 and May 2001, since workers and managers were too overloaded to participate during the summer season.[2] To study the effects of the outsourcing of laundry services, we inter-viewed the hotel manager and the head of housekeeping and observed room cleaners' manipulation of defective linen. Moreover, we asked 18 day-shift room cleaners to put aside for us the unusable bed linen found during the course of a single day in Hotel South, which we then examined for tears, spots, and signs of wear.

We periodically presented our results, first to those cleaners whom we had observed, then to the entire group of cleaners at the hotel concerned, and finally to the study committee. At each presentation, we incorporated into the report feedback and suggestions for further research, together with proposals for changing how cleaning work is done. We gave the final reports to the participating hotels and to the cleaners and the central labour federations to which the unions were affiliated. Based upon these studies, the unions representing workers in this sector were able to negotiate improvements in working condi-tions and, more broadly, to gain recognition for the value and difficulty of these women's work. The results of the first intervention at Hotel East were widely publicized. The Confédération des syndicats natio-naux (CSN), a major union federation, organized a colloquium to use the results to inform and mobilize workers for their upcoming contract negotiations. One hundred and fifty cleaners from 80 hotels came to the colloquium. The presentations, as well as the discussions among cleaners from different hotels, favoured negotiating changes in work-load and organization. Subsequently, the union developed contract langu-age concerning cleaners' workloads, whilst the cleaners themselves were delighted at being involved and having their situation well understood.

On the Logistics of Cleaning

When we conducted our study, cleaners at Hotel East represented 20% of all workers at the hotel. Almost all were women, of whom 69% were immigrants, from ten different countries. Their average age was 50, ranging from 25 to 67, and their average seniority was 15 years (range 1–30 years). In Hotel South, room cleaners were one quarter of all hotel workers (79% were immigrants, from 18 different countries), with their average age being 42 (19–67) and average seniority 10 years (3 months–24 years). In both hotels the largest single group of workers was that composed of white women of European or North American origin, though other substantial groups of cleaners were born in Haiti, Chile, and Jamaica. The majority of cleaners worked in cleaning services permanently, although some of them were students employed only during summer time.

Work in hotels is seasonal, with occupancy at its maximum (about 80%) in summer. During the rest of the year the rate of occupancy ranges from 50% to 77% (Statistics Canada 2000). Typically, employers adjust both the numbers of employees and the hours of work per employee as a function of demand, with 63% of Québec hotel workers employed either part-time or seasonally (Conseil québécois de ressources humaines en tourisme 1999:78). Hours of part-timers fluctuate from 17 to 29 hours per week (Conseil québécois de ressources humaines en tourisme 2001:2). According to employer sources, major hotels in Vancouver, Toronto, and Montréal are 80% unionized (Stokes 1997). In the hotels we studied, one union represented all employees. Cleaners can have regular or casual status and work hours are distributed among all workers by seniority, according to the collective agreement. In Hotel East it takes 15 years' seniority to reach full-time, full-year status (48% of employees) and in Hotel South it takes 17 years (39% of employees).

Room Assignments

In Hotel East, each cleaner was assigned 15 rooms, with a reduction if she had to do more than one "VIP" room (expensive rooms which require cleaning to a higher standard and contain more items to be cleaned). After the results of the first study, the union negotiated a reduction in the number of rooms to be cleaned, such that by the time we studied Hotel South, cleaners in both hotels were assigned 14 rooms, with reductions for VIP rooms or for rooms located on different floors or sections. The weekly schedules were usually made known on Thursday for the week beginning on the following Sunday. However, if occupancy rose unexpectedly, some workers were solicited for extra hours for the following day or even the same day. Less-senior cleaners' assignments could change from day to day, and they were more often allocated rooms spread over several floors or sections.

Work Activity

Cleaners generally work alone. In a typical workday, one hour will be spent getting room assignments, retrieving clean linen, emptying soiled linen, and preparing their supply cart for the next day. The average time allotted for cleaning each room rose from 26 minutes in the first study to 28 minutes in the second. Our observations show that the time to clean a room usually varies from 15 to 49 minutes, but can reach 90 minutes in exceptional situations. Work is interrupted by two 15-minute breaks and a lunch period of 1 hour.

Cleaners' work activity varies significantly with the type of room to be cleaned and according to whether or not the guest is staying on (see Table 1). For example, newly-vacated rooms have more stringent

Table 1: Description of operations performed to clean an occupied room and a check-out room

Task	Operations in a room where occupants are staying on	Additional operations when occupants are checking out
Stripping the room	Collect trash and empty containers	Wash trash containers
	Put away ironing board and iron	
	Take out dirty bed sheets, pillows and towels	
	If dirty, wash dishes, ice and coffee buckets, trays, ashtrays, etc	Even if not dirty, wash trays and ice and coffee buckets
Making the bed	Change two sheets, tidy blanket and bedspread	If necessary, change bedspread
	Put pillowcases on three pillows	
Cleaning the bathroom	Dry washed dishes in the bathroom and return them to tray	
	Wash and dry sink, bathtub, toilet bowl and mirrors	Change shower curtains
	Check if any cosmetic products (shampoo, conditioner, bath soap, hand soap, shower cap, shoe cleaner) are missing and replace them	Wash cosmetic tray
	Replace towels	
	If necessary, replace tissues and toilet paper	
Dusting, vacuuming and finishing the room	Dust bedside table, tables, head of bed, commode surface, television front and upper surface, mirrors	Dust bedside table and commode drawers, back of television, lamps, heaters, chairs and footstools, curtains, picture frames
	Vacuum floors	Vacuum under the cushions of sofas, behind furniture, under bed
	Inspect closet Close the door	
	Push cart to the next room	

standards for dusting, toilet cleaning, and dishwashing. Dusting time varies from 1 minute for a room that will be re-occupied by the same guest to 6 minutes if the guest is leaving. Similarly, dishwashing time typically runs from 2 to 8 minutes.

Factors Intensifying the Workload

Cleaners report that their workload has intensified over the last 10 years as a result of several factors. First, "spring cleaning" has been reassigned to the daily routine. Whereas in the early 1990s cleaners did such major housekeeping tasks as turning mattresses and cleaning fans, electric heaters, and corridors during periods of low occupancy, these tasks are now supposed to be part of the daily routine and no specific time or personnel has been assigned to them. Second, organized tour groups now arrive more frequently—the result of hotel marketing strategies geared towards increasing occupancy rates—and this means that many rooms are vacated at the same time. Third, as mentioned previously, a number of new amenities introduced due to marketing initiatives must be replaced, dusted and/or cleaned, usually daily. Other marketing devices have been introduced from time to time, with dramatic consequences for cleaners. In Hotel East, for example, families with children received a game with tiny pieces, with the result that, on departure, cleaners had to check every corner of the room, under cushions, and in drawers to make sure the pieces were all accounted for. A fourth intensification concerns bedding. Specifically, mattresses and bed linen have been replaced by bigger and heavier versions. Hotels have introduced more king-sized beds which typically take approximately 1.6 times as long to make as double beds. Cleaners complain particularly about the heavier mattresses because they have to lift them as many as eight times to introduce the sheets. Furthermore, the bigger sheets which these beds require are harder to introduce under the mattresses and require the mattress to be lifted higher. In addition, as part of an effort to make beds look all white and clean when the bedspread is turned back a third sheet has been added. Although this serves no purpose other than an aesthetic one, this practice has been extended to all beds in both hotels, multiplying the time to make a bed by a factor of 2.4.

Fifth, new decorations have introduced more hard-to-reach mirrors, porous surfaces that pick up stains more readily (such as synthetic marble in bathrooms and coffee tables), and dark furniture that is more exacting to clean. Finally, laundry has been outsourced to cut costs. However, this has produced additional work for the cleaners. Thus, linen being returned from the laundry services was often missing or late—on average, from September to May the cleaners had to wait for linen to arrive from the outside one day per week, with the situation being much worse during the busy summer season. Such delays resulted in cleaners having to make rooms in two steps rather than one. Equally, according to the cleaners there was much more damaged linen after laundering was contracted out and this added to the cleaners' work—a cleaner would start to make a bed with a sheet and then notice it was torn or stained and, whereas she would sometimes be able to conceal the damaged area by putting it toward the bottom of the bed, often she would discover a large spot or tear toward the end of bed-making and so would have to start again from scratch. These problems were compounded by the fact that it was no one's job to inspect the laundry before returning it to the contractor. Thus, damaged items were tossed back into the laundry, washed again, and put back into circulation. The proportion of damaged items thus rose over time, making it harder and harder for the cleaners to find enough good sheets and increasing the time for room cleaning. In the case of 252 beds from 196 rooms we observed, 26% had a damaged sheet and 34% had a damaged pillowcase.

The above changes affected all cleaners, but less-senior workers faced even more stringent conditions for three principal reasons: (i) they have to travel longer distances to reach their assigned rooms because contiguous rooms are generally assigned to cleaners with more seniority; (ii) they are more likely to run short of supplies such as clean linen because they more frequently work afternoon shifts, by which time supplies have already been monopolized by day-shift cleaners; and (iii) because they work afternoon shifts they are more likely to spend time looking for a free cart or vacuum cleaner. The magnitude of these problems is highlighted by the fact that on at least one occasion our observations showed that, on the afternoon shift at Hotel South, eight of ten storage areas lacked a vacuum cleaner and three lacked a cart. We observed one search that lasted over an hour whereas in another case the only cart found was loaded with soiled linen and waste and took 45 minutes to clear and load properly. Significantly, as a way to avoid this situation, some afternoon shift cleaners actually hid favourite carts and vacuum cleaners so as to be sure to find them when they came back to work, which was often several days later. This strategy, of course, may be helpful to an individual cleaner but it has significant drawbacks for her colleagues.

Despite these difficulties, room cleaners generally forced themselves to clean thoroughly all the rooms assigned to them, although this was not required by their contract. In interviews, cleaners explained this by saying that they wanted to be proud of their work but, also, that they were afraid of being criticized by the head of housekeeping, criticism which might result in being allocated fewer hours of work. Less-senior workers were particularly worried in this regard. Although the union's contract provided that housekeeping

would call cleaners to work in order of seniority and that time would
be allowed for call backs, many worried that if they were perceived
as poor workers housekeeping would be less patient in waiting for
them to call back. They were also afraid of getting poor room assignments in terms of locations or types of rooms.

As a way to finish all their assigned rooms, cleaners regulated their
workload in a number of ways. A commonly-used strategy was to
anticipate some operations. Hence, the day before she anticipated
having to clean a large number of rooms belonging to guests who
would be checking out, a cleaner would usually replace the toilet
paper rolls and the tissue boxes, vacuum behind the furniture, and
change bedspreads and shower curtains. Another strategy was to dis-
tribute tasks over several workdays. Thus, deep cleaning one room at
a time over a long period would enable a cleaner to clean more rapidly
on days when the workload was heavy. However, these strategies
are only available to the minority of cleaners who are senior enough
to have their own regularly assigned territory. Cleaners without an
assigned territory, by contrast, could generally save time only by
rushing, skipping breaks, or leaving out parts of the job (usually the
cleaning of picture frames, lamp bases, under beds, mirrors, dishes,
and toilets), a strategy which then left them open to charges of
laxity—according to the inspection sheets from 30 rooms in Hotel
East, for instance, the cleaning of 23% of rooms was judged to be poor
or unacceptable. Another strategy was to access help from friends,
usually those from the same national and language group, although
such help often had to be kept secret to protect the friend from being
fired and was possible only if the helper herself was not overloaded.

Union Resistance

Whereas individualized strategies such as those just mentioned were
one way in which cleaners sought to negotiate the new work regime
implemented in the two hotels we studied, the involvement of the
Confédération des syndicats nationaux (CSN) labour federation
represented a more collectivized response. Given that the CSN had
been involved in developing the study outlined above, it is, perhaps,
no surprise that it has been intimately involved in seeking to help hotel
cleaners resist some of the efforts at work intensification pursued by
management. Consequently, after the results of the Hotel East study
were published during the summer of 2000, workers put their case
to the public through demonstrations and leafleting. Negotiations
with employers were eventually successful in reducing the numbers
of rooms that had to be cleaned in a shift (from 15 to 14) and in
introducing various mechanisms for lightening cleaners' workload
when large numbers of clients arrive or leave at the same time. These

provisions were later extended to almost all of the unionized hotels in Montréal. For its part, the study at Hotel South resulted in further recommendations which were presented to the employer and the union in 2002. These results were used to gain more improvements in working conditions, such as taking into account the number of beds and occupants per room when assigning rooms to clean. Additional gains in all hotels in Montréal were attained through a short strike in the summer of 2005.

Clearly, the CSN is not alone in seeking to improve cleaners' lot, as witnessed by the Justice for Janitors campaign in the United States and a recent strike by cleaners in Paris (*L'Humanité* 2005). However, what is important with regard to the CSN's activities in Québec is that the movement for change has been reinforced by a tradition of struggle in the union involved, as well as the university–union collaboration outlined above. Indeed, the fact that the unions partnered with scientists to produce quantitative data on the cleaners' workload was helpful in building support for the cleaners' struggle within the local union and within the union federation.

Although the CSN, then, is certainly aware of the consequences of the kinds of precarious work that have emerged in the hotel industry and has tried to protect less-senior workers by, for example, negotiating access to temporary or part-time jobs as waitresses in other unionized areas, it realizes that these individual *ad hoc* solutions do not provide much long-term economic security to cleaners. Consequently, the union has sought to develop contract language that will protect cleaners and has also tried to encourage greater state regulation of the industry. However, there are limits to the powers of provincial and even national unions in a context of governmental deregulation and the increasingly cut-throat corporate competition which has developed as hotel chains have had to satisfy the short-term demands of global stock markets. Equally, the growing power of the hotel chains which has resulted from acquisitions and mergers means that, even if the state wished to do so, its ability to set and enforce standards has become increasingly difficult. Nevertheless, just because something is difficult does not mean that it should not be attempted, and unions in Canada and elsewhere have pursued ergonomic regulations as part of their struggles to improve workers' conditions of employment.

Regulations and Standards

Those doing business in a globalized marketplace have an interest in setting international standards for products, such that, for example, a television made in India can be plugged into a socket in Cameroon. To facilitate this the International Organization for Standardization

("ISO") has promulgated a wide range of standards for a large number of commodities and processes, from the procedures for testing dental amalgams (ISO 10271 2001) to the dimensions and tolerances of watch glasses (ISO 14368–1 2000). ISO standards also include what are called ergonomic norms (the ISO 9000 series), specifying such things as limits for weights lifted and parameters for office equipment. This is important, because although they do not have the force of law, a company that can boast that it has been certified to conform to ISO standards may enjoy a competitive advantage.

On the face of them, these standards are potentially a good thing for workers since they provide an opportunity to improve working conditions. However, there are several problems with them. First, a company that has been certified may not in fact conform to the standard since there is little inspection following certification (Laperrière 2004; Toulouse 2003). Second, the standards are extremely precise, to the point where they may not apply in real life. For example, the standard for weight-lifting (ISO 11228–1) specifies the distance between the centre of mass of the worker and the centre of mass of the object to be lifted. Yet these standards do not take into account differences among individual workers (eg, the back of a female worker with large breasts has a very different biomechanical relationship to the weight lifted from the back of the average male worker [Tate 2004]). Equally, the specifics of the workers' situation may not have been considered. Thus, lifting a wriggling child does not have the same effect on the human body as does lifting an inert object of the same weight. Thirdly, each standard takes a long time to develop and covers only a tiny proportion of actual risky situations. It is therefore impossible to describe in detail every working posture that is associated with a health risk, for the number of working postures is nearly infinite—for instance, struggling with the mattress of a king-sized bed involves combinations of postures that vary with the worker and the specifics of the hotel room. In practice, then, it would be very hard for an international organization to promulgate prescriptive standards that are enforceable and which take the whole range of workers' real-life situations into account.

Equally, hotels are constantly producing environmental changes that affect posture, generally at a rate far outpacing the ability of regulatory entities to set standards. Moreover, hotels are being concentrated into chains with distant headquarters and are "branded" so that guests might expect the same levels of service and amenities no matter at which of the chain's hotels they actually stay. Accordingly, even if they might wish to be more responsive to the needs of their workers, local managers usually must follow corporate directives that often extend to the details of furniture, decoration, tools, and even which cleaning products to use. Thus, there has been a proliferation of mirrors which require careful daily cleaning but which are positioned according to corporate specifications, often at the limit of cleaners' reach—a situation which forces them to adopt awkward postures. Although the union complains that such insidious changes, over which it has no control, counter the gains they made when using the research outlined above in contract negotiations, at this point there is no venue where parameters that concretely influence workload could be regulated at a supranational level. Consequently, only local negotiations can do this, and only where a strong union movement exists.

Nevertheless, despite such problems we think that it is vital to support the development of ergonomic requirements that are based on the general promotion of health rather than specified, limited operations (Lippel and Caron 2004). This type of regulation is currently in force in British Columbia, and is being developed in Ontario. While waiting for such regulations, local cleaners' unions and associations must be supported in their struggles for recognition of the difficulties and the importance of their work, so that those in the boardroom will become more aware of the costs of their marketing and employment strategies for the performance of cleaning. Concretely, this means that legislation and regulations supporting union organizing should also be supported.

Discussion

Changes associated with neoliberal globalization, such as cuts in work hours, outsourcing, restructured tasks, added amenities, and the standardization and upscaling of hotel furnishings, have intensified work and diminished cleaners' ability to regulate their workload. All cleaners suffer the consequences of these changes, but the impact is worst for the least senior. Speed-up and a growing precariousness of work have also caused social problems amongst cleaners, some of whose strategies for self-protection work to the detriment of other cleaners. Thus the growing precariousness of work in the industry encourages workers to vie with each other for the favours of the housekeeping administrators who allocate work and hours, an outcome which increases the power held over them by their employers.

Along with these changes, the transnational migration of cleaners has led to a diversification of the racial/ethnic and class composition of this population, with resulting contradictory effects: solidarity within groups and classes but also, often, conflict between them. Hence, cleavages repeatedly emerged amongst the cleaners we studied, based upon seniority (those who had a designated territory versus those without), on status (casual student cleaner versus regular cleaner), and on nationality (Canadian-born versus immigrant or, to a lesser extent, among immigrant groups). Regularly, less-senior cleaners accused regular cleaners of hiding linen and doing a poor job on "their" rooms the day before a day off, leaving more work for their replacements. In turn, the regular cleaners complained that those who replaced them often failed to load the linen carts properly, to roll up the vacuum cleaner cord, or to change vacuum cleaner bags. Significantly, conflicts based on nationality were intermeshed with those based on status, arising both amongst cleaners and between cleaners and administrators. Thus, in Hotel South immigrant groups perceived less-senior, white, Canadian-born cleaners, especially (but not only) summer students, as lacking solidarity, as more prone to yield to management demands, and as contemptuous toward immigrants with poorer language skills. By contrast, senior, white, Canadian-born cleaners viewed some educated Latin American immigrants as aloof for insisting on receiving recognition for educational or professional status they had attained in their home country. These situations were exacerbated by the fact that "favours" done by administrators were often interpreted as arising from national or ethnic preference—for example, in Hotel South a housekeeping employee was suspected of assigning the "best" rooms to the cleaners of her ethnic group, and in Hotel East the supervisors were thought to be easier on their compatriots.

As competition intensifies, hotels will undoubtedly continue to change labour practices and add to the amenities they offer as a way both to try to minimize costs and to attract more high-end clients who provide the largest profit margins. Since cleaning is far from the minds of those who devise these work and marketing strategies, any changes in products and services are likely to have major unforeseen consequences for cleaners' workloads. Moreover, public awareness of the health and safety implications for cleaners of workplace design is even lower than it is for factory workers, because the hotel workplace is perceived primarily as a client-service area. Certainly, then, we would urge that local cleaners' unions and associations be supported in their struggles for recognition of the difficulties and the importance of their work, such that those in the boardroom—and, more importantly, the general public who might bring pressure to bear on hotel management—will become more aware of the costs of marketing and employment practices for the performance of cleaning. At the same time, whereas employers often simply assume that work restructuring through intensification and outsourcing will save them money, our research has shown that when these strategies are applied without understanding the concrete realities of cleaners' work, there may often be, in fact, significant losses for employers, both in terms of economics and of quality. Employers should be made more aware of such costs.

Acknowledgments

We thank the workers involved in the study and their union, the Commerce Federation of the Confederation of National Trade Unions. Part of the funding for the studies was negotiated through the union and provided by one of the hotels. The authors are part of a partnership with the three major Québec union centrals, supported by the Fonds Québécois de Recherche sur la Société et la Culture. We thank Suzy Ngomo for technical help.

Endnotes

[1] The hotel names are pseudonyms.

[2] We should point out, then, that because data were collected during the winter, when work was not always available for them, we were not able to reach many of the less senior cleaners. Consequently, our results may over-represent the experience and perceptions of more-senior, regularly employed cleaners. This means that what we report here represents the best face of the industry and that working conditions for cleaners overall may very likely be worse.

References

Bernhardt A, Dresser L and Hatton E (2003) The coffee pot wars: Unions and firm restructuring in the hotel industry. In E Appelbaum, A Bernhardt and R Murnane (eds) *Low Wage America: How Employers Are Reshaping Opportunity in the Workplace* (pp 33–76). New York: Russell Sage Foundation

Cline R S (nd) Hospitality adjusts to globalization. Arthur Andersen's Hospitality Consulting Services, New York. http://www.hotel-online.com/Trends/Andersen/global.html (last accessed 26 January 2006)

Conseil québécois de ressources humaines en tourisme (2001) Enquête portant sur la rémunération des employés de l'industrie touristique (2001). Partie 1: Profil des entreprises participantes et de leurs ressources humaines. http://www.cqrht.qc.ca/CQRHTWeb/fr/public/commun/publications/contenu/documents/remuneration1.pdf (last accessed 1 April 2005)

Conseil québécois de ressources humaines en tourisme (1999) Mise à jour 1999 des données du diagnostic d'ensemble des ressources humaines en tourisme. http://www.cqrht.qc.ca/CQRHTWeb/fr/public/commun/publications/contenu/documents/diagnostic_1999.pdf (last accessed 1 April 2005)

Cortina L M, Magley V J, Williams J H and Langhout R D (2001) Incivility in the workplace: Incidence and impact. *Journal of Occupational Health Psychology* 6(1):64–80

Dejours C (1993) *Travail: Usure Mentale.* 2nd ed. Paris: Bayard

Gaucher D (1981) L'égalité salariale des femmes: Ebauche d'une problématique de la discrimination sexuelle et quelques données. In C Bernier, R Bibeau, J Dony and P Doray (eds) *Travailler au Québec* (pp 183–197). Montréal: Albert St-Martin

Glenn E N (1992) From servitude to service work: Historical continuities in the racial division of paid reproductive labor. *Signs* 18(1):1–43

Glenn E N (2001) Gender, race and the organisation of reproductive labor. In R Baldoz, C Koeber and P Kraft (eds) *The Critical Study of Work: Labor, Technology and Global production* (pp 71–82). Philadelphia: Temple University Press

Guérin F, Laville A, Daniellou F, Duraffourg J and Kerguelen A (1997) *Comprendre le travail pour le transformer.* Montrouge, France: ANACT

Karjalainen A, Martikainen R, Karjalainen J, Klaukka T and Kurppa K (2002) Excess incidence of asthma among Finnish cleaners employed in different industries. *European Respiratory Journal* 19(1):90–95

Laperrière É (2004) "La posture debout prolongée en milieu de travail: différences à court terme des multiples types de posture debout." Master's thesis, Department of Biological Sciences, Université du Québec à Montréal

Leary M R and Kowalski R M (1995) *Social Anxiety*. New York: Guilford Press

Lee P T and Krause N (2002) The impact of a worker health study on working conditions. *Journal of Public Health Policy* 23(3):268–285

L'Humanité (2005) Grève dans un quatre-étoiles. 16 February. http://www.humanite.fr/journal/2005-02-16/2005-02-16-456771 (last accessed 29 January 2006)

Lippel K and Caron J (2004) L'ergonomie et la réglementation de la prévention des lésions professionnelles en Amérique du Nord. *Relations Industrielles/Industrial Relations* 59(2):235–272

Messing K (1998) Hospital trash: Cleaners speak of their role in disease prevention. *Medical Anthropology Quarterly* 12(2):168–187

Messing K, Chatigny C and Courville J (1998) "Light" and "heavy" work: An analysis of housekeeping in a hospital. *Applied Ergonomics* 29(6):451–459

Messing K, Haëntjens C and Doniol-Show G (1993) L'invisible nécessaire: L'activité de nettoyage des toilettes sur les trains de voyageurs en gare. *Le Travail Humain* (55):353–370

Messing K and Seifert A M (2001) Listening to women: Action-oriented research in ergonomics. *Arbete och Hälsa* 17:93–104

Messing K, Seifert A M, Vézina N, Balka E and Chatigny C (2005) Qualitative research using numbers: An approach developed in France and used to transform work in North America. *New Solutions: A Journal of Occupational and Environmental Health Policy* 15(3):245–260

Milburn P D and Barret R S (1999) Lumbosacral loads in bed making. *Applied Ergonomics* 30(3):263–273

Muqa F, Mathieu F, Sanchez M I, Roux F and Crestois M (1996) La femme de chambre. *Cahiers de Médecine interprofessionnelle* 3:305–316

Prévost J and Messing K (2001) Stratégies de conciliation d'un horaire de travail variable avec des responsabilités familiales. *Le travail humain* 64:119–143

Quinlan M, Mayhew C and Bohle P (2001) The global expansion of precarious employment, work disorganisation, and consequences for occupational health: A review of recent research. *International Journal of Health Services* 31(2): 335–414

Rosenman K D, Reilly M J, Schill D P, Valiante D, Flattery J, Harrison R, Reinisch F, Pechter E, Davis L, Tumpowsky C M and Filios M (2003) Cleaning products and work-related asthma. *Journal of Occupational and Environmental Medicine* 45(5):556–563

Scherzer T (2003) The race and class of women's work: Reproducing inequalities in hospital nursing. *Race, Gender and Class* 10(3):23–41

Scherzer T, Rugulies R and Krause N (2005) Work-related pain, injury, and barriers to workers' compensation among Las Vegas hotel room cleaners. *American Journal of Public Health* 95(3):483–488

Seifert A M (2001) *Intervention Ergonomique sur le Travail des Préposées aux Chambres*. Montréal: CINBIOSE, Université du Québec à Montréal

Seifert A M and Messing K (2004) Looking and listening in a technical world: Effects of discontinuity in work schedules on nurses' work activity. *Perspectives Interdisciplinaires sur le Travail et la Santé* 6(1). http://www.pistes.uqam.ca/v6n1/articles/v6n1a3a.htm (last accessed 19 December 2005)

Seifert A M, Messing K and Dumais L (1997) Star wars and strategic defense initiatives: Work activity and health symptoms of unionized bank tellers during work reorganization. *International Journal of Health Services* 27(3): 455–477

Statistics Canada (2000) *Caractéristiques et performance des hôtels et des hôtels motels.* Série d'études analytiques No 33. Indicateurs des services. Catalogue No 63–016-XPB

Stokes D (1997) "Bargaining power." *Hotelier* (Toronto) March–April:27–30

Tate A J (2004) Some limitations in occupational biomechanics modelling of females. Communication 29 March 2003 given at the Colloquium on Women's Environmental and Occupational Health, Montréal. Available on CD through the Women's Health Bureau of Health Canada, Ottawa

Teiger C and Bernier C (1992) Ergonomic analysis of work activity of data entry clerks in the computerized service sector can reveal unrecognized skills. *Women Health* 18(3):67–77

Torgen M, Nygard C H and Kilbom A (1995) Physical work load, physical capacity and strain among elderly female aides in home-care service. *European Journal of Applied Physiology and Occupational Physiology* 71(5):444–452

Toulouse G (2003) *L'intégration de la prévention SST au management des normes ISO 9000 par la macroergonomie: Une recension des écrits suivie d'un projet pilote.* Montréal: Institut de recherche Robert-Sauve en santé et en sécurité du travail. Report R-324

Wilkinson R G (1999) Health, hierarchy, and social anxiety. *Annals of the New York Academy of Sciences* 896:48–63

Chapter 9

Work Design and the Labouring Body: Examining the Impacts of Work Organization on Danish Cleaners' Health

Karen Søgaard, Anne Katrine Blangsted, Andrew Herod and Lotte Finsen

Introduction

In the European Union (EU), it is estimated that private enterprises, governments, and local authorities employ nearly three million full- and part-time cleaners (95% of whom are women) (Krüger et al 1997). This makes professional cleaning one of the EU's most common occupations. Such ubiquity is significant because although it is generally not perceived as being a hazardous job in the same way as is, say, working in manufacturing, cleaning can, in fact, place a considerable strain on the labouring body, especially with regard to cleaners' cardiovascular and musculoskeletal systems and their mental health (Nielsen 1995; Sjögren et al 2003). Thus, a recent survey amongst 2270 Danish full-time cleaners showed that 20% suffered daily pain as a result of their work (Bonde and Jensen 2004), whilst another survey demonstrated that cleaners experienced some of the highest degrees of monotonous repetitive work with little variation of any profession (Borg and Burr 1997). As a consequence of such working conditions, many cleaners are forced either to opt for early retirement or are, essentially, invalided out of the profession (Borg and Burr 1997), a phenomenon with sizeable consequences both for themselves individually and for society more broadly, which must pay the healthcare and other costs associated with their work injuries.

Given its low social status, for many years professional cleaning in Denmark was seen as an unskilled job which could be done by pretty much anyone. As a consequence, there was little incentive on the part of employers to mechanize or redesign cleaning work to improve efficiencies. During the last 20 years or so, however, considerable changes have been introduced into the industry and, so, into cleaners' work lives (Aguiar 2001). Specifically, in Denmark as elsewhere, privatization and a spate of acquisitions and mergers facilitated by market deregulation have led to both a centralization of ownership and a substantial specialization of cleaning work as companies have focused upon particular economic sectors (offices versus factories, healthcare facilities versus banks), with the result that firms such as the Danish cleaning giant ISS now typically market themselves as having specific expertise in a host of arenas, including the hospital, washroom, food processing, computer server, copy machine, and automotive industries, all of which may require quite different and unique sets of cleaning skills.[1] At the same time, efforts to facilitate work efficiency so as to increase profitability and thus to please shareholders and the market have led to the introduction of new technologies, such as microfibre mops and cloths (United States Environmental Protection Agency 2002), triangular handles on floor machines which allow easier gripping and so faster work, and newer and more powerful cleaning solvents (Sanders 2001).

Given the potential implications for cleaners' health and worklife of such specialization and intensification, in 1995 the European Commission (in the form of the Biomedical and Health Research Programme—BIOMED2) initiated a four-year, multidisciplinary research and development network entitled "Risk in Cleaning—Prevention of Health and Safety Risks in Professional Cleaning and the Work Environment". Experts on the occupational hazards of cleaning from Germany, Finland, Denmark, and Italy participated in the project, the goal of which was to synthesize results from a series of national studies so as to provide a basis for developing measures to improve health and safety conditions for cleaners. One of the outcomes of this project was a literature-based, state-of-the-art report which highlighted the fact that cleaners' jobs involve significant repetitive tasks (Krüger et al 1997). On the strength of this report, in 2001 the Danish Minister of Labour acknowledged that changes in the industry's structure were bringing about a deterioration of cleaners' work environment and suggested that increased task variation, job enlargement, and greater job flexibility might be encouraged so as to rectify the situation. Consequently, the Danish Working Environment Authority (WEA)—an agency of the Ministry of Employment tasked with creating safe and sound working conditions at Danish workplaces by conducting inspections of companies, by drawing up rules on health and safety at work, and by providing information to workers and others on workplace health and safety, and with the authority to penalize enterprises not in compliance with working environment rules—funded a number of projects designed to improve women's working conditions in general, and to investigate specifically the potential benefits of organizing cleaning work in different ways.

Drawing upon ergonomic research in other industries which shows that the repeated performance of identical tasks can have deleterious effects upon the human body and that varying such tasks and providing workers with frequent breaks from those tasks has beneficial outcomes, the specific research questions the projects posed focused upon whether the differences between how cleaning work is currently organized and how the WEA would like it to be organized (such as encouraging job enlargement) would actually be sufficient to bring about a healthier workplace, at least from an ergonomic perspective. As might be imagined, such questions are politically quite contentious. Thus, although government workplace regulators see encouraging or even mandating larger numbers of breaks and a greater variety of work tasks as a way to develop effective ergonomic standards for cleaning, as outsourcing has encouraged work intensification the potential for securing recuperative breaks for cleaners during the workday has, in fact, been severely diminished as they have generally faced a steady increase in the amount of space they must clean and the pace at which they must do it (Bonde and Jensen 2004). Likewise, although the emergence of large conglomerates such as ISS who operate in several different sectors would appear, in theory at least, to offer possibilities for job enlargement and task variation, as companies specialize within particular sectors and as the internal lines of divisional organization harden over time within these conglomerates, the possibilities for variation and enlargement in work activity are actually often reduced—workers who clean offices may not have the specialized knowledge to clean, say, laboratories, so that, generally, the only variation over a workday that can be introduced in their jobs is to switch between the main tasks of floor cleaning and of cleaning surfaces other than the floor, such as furniture (what we are here calling "surface cleaning"). Furthermore, given that, ergonomically speaking, such variation is best if conducted within a single work day—rather than working one job for several weeks and then switching to another—opportunities to shift periodically between divisions may not be all that helpful for cleaners' bodily health, even if they may help alleviate some issues of boredom. Put another way, the restructuring of the industry appears to be at odds with what is beneficial ergonomically for the labouring body.

Certainly, then, in considering different ways to improve workplace health and safety the Danish government has a visceral recognition that interspacing cleaning tasks with other activities and rest may offer some variation in cleaners' cardiovascular and musculoskeletal load during the workday, and a degree of job enlargement has already been introduced into the healthcare sector (specifically, in a number of hospitals) by having cleaners conduct some kitchen and portering tasks. However, as Dempsey and Mathiassen (2006:35) have recently pointed out, such job rotations are frequently "guided by instinct rather than science". Consequently, for effective prevention of workplace injuries over both the short- and the long-term it is necessary, we believe, to identify empirically those work conditions which represent a biomechanical and physiological overload for different parts of the labouring body. It is within this context, then, that this paper seeks to provide an ergonomic analysis of efforts to design an optimal workday using the strategy either of increasing the degree of work variation amongst existing cleaning tasks or the strategy of adding non-cleaning tasks to the cleaners' workday. The data presented here are based upon a literature review of reports produced as part of the EU project mentioned above, a laboratory study which analysed the load on the shoulder muscles in different cleaning tasks and, finally, a field study which focused on the physiological and psychosocial effects of job enlargement.

Our paper is organized as follows. First, we provide a brief overview of the ergonomic literature as it relates to cleaning. We then provide an ergonomic analysis of the impacts of varying cleaning work through the introduction of new equipment and through having cleaners switch between several different ways of cleaning. Third, we explore how job enlargement practices may impact the human body at work. Finally, we draw some brief conclusions about how cleaning should be redesigned if effective ergonomic standards are to be developed.

A Brief Literature Survey
Floor Cleaning

Traditionally, cleaning tasks have mainly been conducted manually and have usually been performed with simple hand-held equipment, such as scrubbing brushes, cloths, and mops. Of late, however, new equipment and methods have been developed which are designed to improve cleaning efficiency and, if the equipment manufacturers are to be believed, also to decrease the physical workload placed on cleaners. Given that floor cleaning is one of the most time-consuming aspects of cleaning, most of the development of new equipment and methods (such as the use of new material such as microfibres, different shapes of handle and mop, as well as dry and damp mopping methods) has focused upon this task, as a way to speed it up.[2] In this context, then, a number of studies have sought to quantify the muscle load—that is to say, the degree of activity in the shoulder during floor cleaning—by estimating the amplitude probability distribution function (APDF) of electromyographic (EMG) recordings (Jonsson 1978).[3] Such studies have generally compared the traditional scrub and cloth method (in which cleaners manually wet and rinse the cloth) with other mopping methods (Winkel et al 1983; Søgaard 1994; Søgaard, Fallentin and Nielsen 1996). Thus, Hagner and Hagberg (1989) have compared cleaners' muscle loads using a "figure-of-eight" mopping method (ie moving the mop in an arc) and using a "push/pull" method (ie moving the mop back and forward) during floor mopping, finding that the static load on the shoulder muscles, with the hand and arm used in the upper position of the handle, was about 10% of maximal voluntary contraction (MVC), regardless of the type of equipment or technique used during floor cleaning—in other words, varying how the cleaning was done made little difference to the pressures placed upon workers' bodies.[4]

Significantly, such muscle load levels are comparable to those found in workplaces with highly repetitive, machine-paced work, such as those involving sewing machine operation and/or assembly work in the electronics industry (Jensen et al 1993; Christensen 1986). However, unlike in these industrial occupations, the median and peak load levels on the shoulder muscles are also high during floor cleaning, a finding which shows that floor cleaning is more strenuous work for the shoulder muscles than those other activities. Similar findings have been reported by Søgaard et al (1996), who analysed signs of muscle fatigue after mopping and recorded a decrease in the mean power frequency (MPF) and a simultaneous increase in the root mean square amplitude (RMS) for the *trapezius* muscle (which connects the back of the skull with the shoulder and the spine) and by Winkel et al (1983), who have also revealed tendencies towards muscle fatigue after floor cleaning.[5] At the same time, studies of cleaners' bodily motions during such floor mopping have shown highly repetitive movements of the upper body extremities, with a cycle time of approximately 2 seconds (Hagner and Hagberg 1989; Søgaard 1994; Winkel et al 1983).

To put this all into non-medical language, then, the EMG signs of fatigue, the findings of the high load levels for the shoulder, and the highly repetitive movements all indicate that floor cleaning is a tremendously strenuous work task for the shoulder muscles, and that these elements may be risk factors in the development of shoulder and neck disorders. Cleaning floors, however, does not affect only the musculoskeletal load on cleaners' bodies. It can also result in a high cardiovascular load. Specifically, when floors are cleaned by mopping, scrubbing or vacuum-cleaning, the stresses on cleaners' hearts and circulatory systems have been observed to increase to between 27% and 55% of their maximal capacity (Hagner and Hagberg 1989; Louhevaara, Hopsu and Søgaard 2000; Søgaard, Fallentin and Nielsen 1996). This is significant because if these levels are compared to the guidelines specified by the International Labour Organisation (Bonjer 1971) and the US National Institute for Occupational Safety and Health (which is part of the Centers for Disease Control) (Davis, Marras and Waters 1998), then a number of the above-mentioned activities demonstrate levels of cardiovascular stress which exceed the guidelines. Moreover, in addition to such ergonomic impacts on the body, jobs that include monotonous repetitive work are often characterized by a poor psychosocial work environment, including few opportunities for mental stimulation, small possibilities for development, and only little social contact and support on the job, all of which can lead to boredom and stress (Borg and Burr 1997; Seifert and Messing, this collection). It is not surprising, then, that a very high frequency of cleaners report poor health and musculoskeletal symptoms, as well as very low levels of *joie de vivre* compared to other employees (Borg and Burr 1997; Burr 1998; Christensen et al 2000; Nielsen 1995; Woods, Buckle and Haisman 1999).

Surface Cleaning

Only one study has assessed the muscle load during surface (ie non-floor) cleaning, which is another frequent task in professional cleaning. Thus, Björkstén et al (1987) measured EMG from the *trapezius*, the *deltoid*, the *extensor carpi radialis brevis*, and the *extensor carpi radialis longus* muscles of the active arm during the cleaning of vertical surfaces.[6] Their research showed a range in the static level from 3% to 6% MVC, in the median level from 9% to 22% MVC, and in the peak level from 38% to 80% MVC for the muscles of the shoulder and forearm—in other words, that this task is as strenuous for the shoulder muscles as is floor cleaning.

With regard to the cardiovascular system, Hopsu (1997) has shown that cleaning vertical surfaces can result in a relative load on the heart of approximately 33% of maximal capacity, a level which is also greater than international guidelines.

Introducing Variation into Cleaning Work Through New Equipment and the Interchange of Cleaning Tasks

Based upon the extant literature in the field, then, and following from the desire of the Working Environment Authority to improve cleaners' working conditions (particularly to reduce musculoskeletal injuries), we conducted a laboratory study to (i) investigate whether new types of equipment for the wet cleaning of floors could reduce and/or change the shoulder's muscle work load; and (ii) to evaluate the load on the shoulder during surface cleaning under different conditions.

Methods

We observed six female cleaners with a mean age of 46 (range 35–57) perform wet floor cleaning with a flat mop and with a wet mop, both of which were long-handled [a detailed description of the applied methods can be found in Søgaard (1994), Søgaard et al (2001) and Laursen, Søgaard and Sjøgaard (2003)].[7] These two methods have been suggested as being less hazardous substitutes for traditional scrub and cloth methods of cleaning. The cleaners employed two different work techniques with regard to the flat mop—either "scrubbing" (moving the flat mop forward and backward in a straight line) or "mopping" (moving the mop in a butterfly motion, from side to side). In the third trial the cleaners used a wet mop, moving it in a butterfly motion. Each trial consisted of one minute of standardized wet floor cleaning (that is to say, the same area of the floor was cleaned and the mops had the same degree of humidity).

The same six cleaners also performed wet surface cleaning of both horizontal and vertical surfaces. We studied four different one-handed operations: (i) cleaning a horizontal surface close to the cleaners (25 cm away); (ii) cleaning a horizontal surface which was somewhat farther away (48 cm) from the cleaners; (iii) cleaning a vertical surface using vertical hand movements; and (iv) cleaning a vertical surface using horizontal hand movements. We recorded surface EMG bilaterally from the *trapezius*, *deltoid*, and *infraspinatus* muscles. The amplitude of the signal was used to express the relative activity level of each muscle. [Signals were full-wave rectified, RMS-converted, and normalized to the maximum EMG amplitude (%EMG$_{max}$) during an MVC for each muscle. Resting level was quadratically subtracted.] An amplitude probability distribution function (APDF) was calculated to give the static, the median, and the peak levels of the load on the shoulder muscles (see note 3 for details on this measure).

Results and Discussion

Floor Cleaning

Data collected from the cleaners' hands and arms when used in the upper position of the mop handle during floor cleaning are presented in Table 1 and compared with those from Søgaard et al (2001), where cleaners used the more traditional minimop (ie a round headed mop with long threads) and the cloth and scrub methods of cleaning. What these data show is that there were significant differences between the more traditional methods of cleaning and the new methods only for the *deltoid* muscle. Despite this apparent improvement for the *deltoid* muscle, however, all the different types of equipment still produced relatively high muscle load levels for the *trapezius*, the *deltoid*, and the *infraspinatus* muscles. The conclusion to be drawn from this

Table 1: Muscular strain in different wet cleaning methods. The values are given as means (range)

Method	n	Muscle	%EMG$_{max}$		
			Static level	**Median level**	**Peak level**
Flat mop—scrubbing	6	Trapezius	8 (2–20)	15 (7–31)	25 (14–49)
Flat mop—mopping			7 (3–17)	18 (10–39)	28 (17–64)
Wet mop			9 (2–20)	17 (9–41)	27 (14–67)
Minimop*			7 (2–14)	13 (7–25)	20 (11–39)
Cloth and scrub*			8 (3–16)	14 (8–25)	21 (12–39)
Flat mop—scrubbing	6	Deltoideus	12 (4–31)	34 (8–89)	55 (16–157)
Flat mop—mopping			8 (3–22)	14 (5–38)	19 (8–49)
Wet mop			9 (3–28)	15 (4–44)	20 (7–57)
Minimop*			8 (2–22)	12 (3–37)	18 (5–50)
Cloth and scrub*			11 (5–21)	28 (8–53)	48 (14–89)
Flat mop—scrubbing	6	Infraspinatus	11 (6–16)	21 (10–30)	34 (14–50)
Flat mop—mopping			8 (3–16)	19 (10–33)	32 (14–66)
Wet mop			10 (3–20)	22 (12–43)	37 (18–88)
Minimop*			11 (3–22)	19 (8–40)	29 (11–67)
Cloth and scrub*			10 (6–14)	18 (9–31)	29 (12–48)

Data for the minimop and the cloth and scrub method from Søgaard et al (2001) are marked with an asterisk (*) and presented for comparison.

analysis, then, is that floor cleaning with more recently developed equipment (such as the flat mop and the wet mop) appears to be just as strenuous for the shoulder muscles as does cleaning using the older tools and methods (ie minimop and scrub and cloth). Changing from older systems to the mopping systems evaluated in this study, then, does not seem to have a beneficial effect on the local load on the shoulder.

Surface Cleaning

Given that surface cleaning is mainly a one-handed task, it has been suggested by some that systematically switching between different cleaning tasks—say, between floor cleaning and vertical surface cleaning—may offer some relief for workers. However, our research shows that such switching does not seem to introduce sufficient variation into the load profile of the shoulder muscles to ameliorate the situation. Thus, during horizontal surface cleaning, the activity in the three muscles of the active arm showed a static level range of 6–8%EMG$_{max}$, a median level range of 10–26%EMG$_{max}$, and a peak level range of 14–47%EMG$_{max}$. The corresponding values for cleaning of vertical surfaces were significantly higher, being 13–29%EMG$_{max}$, 21–48%EMG$_{max}$, and 31–69%EMG$_{max}$, respectively, values which are similar to the results of a study by Björkstén et al (1987). What these numbers show, then, is that surface cleaning is associated with high levels of shoulder muscle load and that, consequently, it does not offer relief from the high loads experienced during floor cleaning. Furthermore, whilst in theory surface cleaning could perhaps allow for some relaxation of the passive shoulder, our results showed this not to be the case and that there was no difference for the shoulder muscles of the passive arm between the four modes of working—all the muscles showed significant contra-lateral co-activation in all four types of cleaning, ranging from 2 to 7%EMG$_{max}$.[8] Therefore, not even the change from a two-handed to a one-handed task seems to offer complete rest for the shoulder muscles of the passive arm.

Introducing Variation in Cleaning Work Through Job Enlargement

Based on these results, we subsequently conducted a field study amongst hospital cleaners to evaluate to what extent switching between cleaning and non-cleaning tasks such as kitchen and/or hospital portering tasks could improve cleaners' work conditions with regard to localized musculoskeletal load, cardiovascular load, and the psychosocial environment. Given the variety of aspects of bodily health in which we were interested, we utilized a number of methods, as outlined below.

Methods

We secured the cooperation of 27 female cleaners from three Danish hospitals (A, B, C), all of whom had at least one year of seniority as cleaners and all of whom volunteered to participate in the study (summary data on the cleaners are presented in Table 2). Some of

Table 2: Mean ± SD age, height, weight, body mass index, seniority, work time, maximal hand grip strength, aerobic capacity, work ability and sense of coherence for the hospital cleaners

Age, years ($n = 27$)	39.3 ± 8.7
Height, cm ($n = 27$)	166.7 ± 6.1
Weight, kg ($n = 27$)	68.7 ± 13.7
Body mass index, kg/m^2 ($n = 27$)	24.8 ± 5.1
Seniority at the workplace, years ($n = 24$)	9.3 ± 7.0
Seniority at the work task, years ($n = 24$)	7.5 ± 7.3
Work time, hours/week ($n = 24$)	36.5 ± 1.6
Maximal hand grip strength, kg ($n = 26$)	36.5 ± 5.3
Aerobic capacity, mlO$_2$/(min*kg body weight) ($n = 27$)	33.0 ± 10.0
Work ability index ($n = 27$)	34.6 ± 3.9
Sense of coherence ($n = 22$)	30.4 ± 6.8

these cleaners predominantly performed cleaning, whilst others also performed kitchen tasks, portering tasks, or both. Hospital C was the only hospital where portering tasks were performed, whilst cleaning and kitchen tasks were performed at all three hospitals. Cleaning tasks consisted of floor mopping, together with the cleaning of furniture and equipment with a cloth. Kitchen tasks included preparing breakfast trays and buffets, making coffee, and dishwashing. Portering tasks primarily included the transportation of beds and wheelchairs (both with and without patients), the transportation of supplies, and the removing of bedclothes. During a typical workday, cleaners would first prepare their cleaning cart and then proceed to their cleaning tasks. Those cleaners performing kitchen tasks generally interrupted their cleaning to help arrange breakfast, lunch, and the afternoon meal. The portering tasks were primarily performed in the morning, though they could sometimes interrupt the cleaning tasks during the day. Work organization was different at the three hospitals. Hence, at hospital A, a supervisor was responsible for operations management at the worksite, whilst cleaners at both hospital B and C were organized into self-operating teams with extended areas of responsibility.

Measurements

All of the cleaners were observed during a full workday and we timed the duration of the six main tasks in which they were engaged, these being: (i) cleaning tasks; (ii) kitchen tasks; (iii) portering tasks; (iv) transportation tasks (exclusive of portering tasks); (v) general "assorted tasks" (eg floor polishing, vacuum cleaning, and cleaning of bed tables); and (vi) meetings. Based on the observations of the time spent on cleaning tasks during the workday, we divided the cleaners into two groups—one was constituted by that half of the cleaners who had the greatest part of their workday occupied by cleaning tasks, whilst the other was made up of the remaining cleaners. As the cleaners undertook their work, we monitored their heart rates. We then compared these for the two groups and for the different work tasks (and pauses). We analysed variation in the heart rate during the workday for the two groups using an APDF where the difference between $P_{0.9}$ and $P_{0.1}$ was used to express the variation.[9] In these APDF analyses, pauses of more than 5 minutes' duration were excluded. We estimated the cleaners' aerobic capacity using a one-point bicycle test (Åstrand 1960; Åstrand and Ryhming 1954).[10] We also used a Jamar dynamometer to measure the maximal handgrip strength of the cleaners' dominant hand, recording the highest value of each of the three gripping attempts each one made.[11]

We then conducted detailed observation of shoulder and back loads for both subgroups during cleaning tasks (shoulder observation $n = 10$;

back observation $n = 5$) and kitchen tasks (shoulder observation $n = 9$; back observation $n = 4$). For the shoulders three categories of load on the dominant shoulder were defined: (i) "heavy load", when the arm was elevated (flexed or abducted) more than 30°, when the shoulder was extended, and/or when the cleaners carried, pushed, or pulled burdens; (ii) "light load", when the arm was elevated (flexed or abducted) less than 30°, and/or when only light burdens (eg cloth) were carried; and (iii) "neutral", when the arm was just hanging down or was fully supported. For the back, two categories of load stress were defined. A "heavy load" was considered to be any instance in which the back was flexed, laterally flexed, or rotated more than 20° with no hand support, or if a heavy object was carried. For an object to be considered heavy, it had to be greater than 11 kg in weight if it was carried close to the body, greater than 7 kg if carried at forearm distance from the body, and greater than 3 kg if carried at three-quarter's arm length from the body.[12] Furthermore, any active use of the mop, together with any pushing or pulling of carts, was considered to constitute a heavy load on the back. By way of contrast, a "light load" on the back was defined as any situation which did not belong to the first category.

All of the cleaners also filled out a questionnaire regarding individual characteristics, physical demands at work, and the psychosocial nature of the work environment, as expressed in terms of the cognitive demands placed upon them (their ability to make decisions, memory demands placed upon them, the ability to show creativity, whether they were given any responsibilities, and so forth) and their possibilities for skill development (what skill demands were placed on them, whether they were provided opportunities to cultivate new skills, whether they were encouraged to take the initiative, and whether they were able to avoid monotony at work).[13] In addition, we analysed cleaners' sense of coherence—that is to say, their global orientation to the world, how they perceive it, and to what extent they find it comprehensible, manageable, and meaningful (Setterlind and Larsson 1995)—as well as their ability to perform their jobs (as expressed in a work ability index; Tuomi et al 1998) and any health or musculoskeletal symptoms they reported (Kuorinka et al 1987). We tested differences between work tasks using the Kruskal–Wallis test, the Mann–Whitney test, and the Friedman test, at a significance level of $P < 0.05$.

Results and Discussion
Cleaner Characteristics
The questionnaire showed that almost all the cleaners were full-time employed and that the average seniority was just under 10 years.

Ninety-six per cent of the cleaners reported they were in good health. The one-year prevalence of musculoskeletal symptoms was as follows: neck 69%, shoulder 56%, arm/elbow 33%, wrist/hand 59%, low back 74%, hip 19%, knee 37%, and feet 41%. These prevalences are at levels similar to those previously reported for cleaners (Borg and Burr 1997; Burr 1998; Christensen et al 2000; Nielsen 1995; Woods, Buckle and Haisman 1999), even though these particular cleaners reported good health relative to cleaners in general. The cleaners' aerobic capacity, maximal handgrip strength, stature and weight (as shown in Table 2) are similar to the mean values for Danish female employees in general (Faber et al 2006; Winkel et al 1983). The cleaners' average work ability as measured on the Work Ability Index (WAI) on a scale of 6–42 (with higher scores indicating better work ability) was 35.[14] Their average score on the sense of coherence index was 30 on a scale of 9–45, wherein a high score has been shown to correlate with positive mental health (Setterlind and Larsson 1995).[15] The observation showed that for the group of 20 cleaners who did kitchen tasks, this task occupied an average of 24% [standard deviation (SD) = 19] of their worktime, whereas for the seven cleaners who did portering tasks this activity occupied 46% (SD = 28) of their time. For the group as a whole, the median value of the observed time spent on cleaning was 72% (SD = 24), whilst 22% of the time spent on cleaning tasks was spent on floor mopping.

Exposure to Repetitive Tasks and Awkward Positions

Almost half of the cleaners reported that they performed floor cleaning during most of the day, whilst more than half of them reported that they performed furniture/equipment cleaning most of the day (see Table 3). Only a few cleaners reported that they mainly performed tasks other than cleaning. Approximately 75% of the cleaners reported engaging in repetitive tasks and 90% of these reported repetitive movements. Experiencing awkward work postures most of the time was reported by 23–36% of the cleaners, depending on the body region, whilst 14% and 19% of cleaners reported lifting and pushing/pulling, respectively.

Heart Rate

As illustrated in Figure 1, cleaners who spent a higher percentage of their workday engaging in cleaning tasks did not have a higher mean heart rate during the workday than the group of cleaners reporting a lower percentage of the workday occupied with cleaning tasks—the rates were 97 beats per minute (SD = 9) for those cleaners whose work involved cleaning tasks during less than 45% of the workday

Table 3: Percentage of cleaners reporting more or less than half the workday spent on floor cleaning, furniture cleaning, equipment cleaning and other tasks, as well as repetitive work and movements and different work postures

	More than half of the time (%)	Less than half of the time or never (%)
Part of work time spent floor cleaning (n = 22)	41	59
Part of work time spent furniture/equipment cleaning (n = 21)	57	43
Part of work time spent on tasks other than cleaning (n = 22)	5	95
Repetition of the same work tasks many times per hour (n = 23)	70	30
Repetition of the same finger, hand or arm movements many times perminute (n = 22)	73	27
Work time with hands above shoulder height or higher (n = 22)	27	73
Work time with the back strongly flexed (n = 22)	36	64
Work time kneeling or squatting (n = 22)	23	77
Work time lifting heavy burdens (n = 22)	14	86
Work time pushing or pulling heavy burdens (n = 21)	19	81

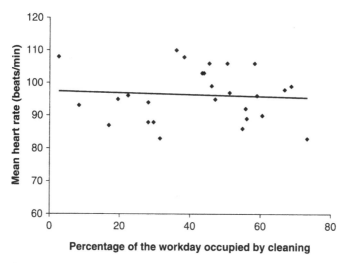

Figure 1: Mean heart rate for the workday of each cleaner ($n = 27$) shown against the percentage of the workday occupied by cleaning tasks. The regression line shows no significant relation between the two parameters.

($n = 14$) and 94 beats per minute (SD = 7) for the group of cleaners ($n = 13$) who cleaned for greater than 46% of their workday. Put another way, there was no association between variations in cleaners' heart rates and the extent of cleaning tasks they did during the workday. Furthermore, we found that there was no relationship between cleaners' heart rates during the workday and their aerobic capacity, meaning that any differences in heart rates during different types of work was not reflecting the cleaners' inherent aerobic capacity but was related, instead, to the physical workloads in which they were engaged at any one time. This is noteworthy because although cleaners' heart rates were measured at a fairly constant average of almost 100 beats per minute during cleaning tasks, kitchen tasks, transportation tasks, and "assorted tasks", they were significantly higher during the performance of portering tasks ($p < 0.05$) and significantly lower during meetings and pauses ($p < 0.05$) (see Figure 2).

The fact that cleaners' heart rates were relatively constant during cleaning, kitchen tasks, transportation tasks, and "assorted tasks", regardless of the amount of time actually spent on these tasks, indicates that these all produced a similar load on the cardiovascular system, which is probably a consequence of the fact that walking and standing dominate virtually all these work tasks. Significantly, such findings are not unique to cleaning. Thus, similarly small degrees of variation have been reported for meat cutters, whose job is also dominated by standing. On the other hand, studies of workers such as garbage collectors, whose work patterns range from sedentary to very active, often in a short space of time, show a much greater heart rate

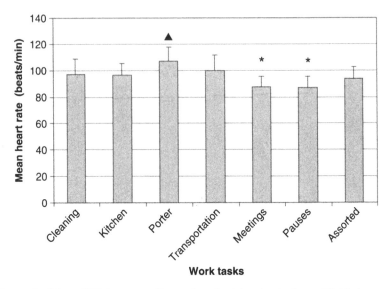

Figure 2: Mean (SD) heart rate for work tasks (cleaning task $n = 27$; kitchen task $n = 22$; porter tasks $n = 9$; transportation $n = 10$; meetings $n = 12$; pauses $n = 26$; assorted tasks $n = 27$). ▲ indicates significantly higher heart rate, whilst * indicates significantly lower heart rate.

variation—as much as 37 beats per minute in one study (Schibye and Christensen 1997). Such findings are noteworthy because research shows that a larger variation in heartbeat is healthier for the cardio-vascular system (Davies and Knibbs 1971).

Based on these results, then, it is necessary to introduce completely different work tasks in cleaners' days if the load on the heart and the circulatory system are to be more varied. Given that cleaning and kitchen and even portering work are all primarily performed standing or walking, including some type of activity in which cleaners sit could potentially provide sufficient physical variation so as to benefit—or, at least, so as not to worsen—their cardiovascular systems. One such work task might be some sort of planning task in a self-organized team, although this would require employers to trust cleaners sufficiently to allow them to engage in such activities, the likelihood of which is questionable, given the current employment trend towards ever greater micro-management and control of the workplace. Given that research suggests that it is the lack of variation in the heart rate and not just the actual heart rate level itself that implies a risk for the cleaners, also introducing short periods of activity that considerably increase the heart rate would be beneficial for cleaners' overall health, although the introduction of new work tasks (such as portering) or other activities such as, perhaps, callisthenics as performed in many East Asian workplaces should not lead to a mean relative heart rate

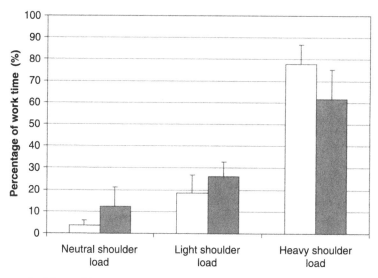

Figure 3: Mean (SD) duration of shoulder load (% of time) is illustrated for cleaning (light bars, *n* = 10) and kitchen work (dark bars, *n* = 5).

that exceeds established guidelines but should, rather, be interrupted by periods of relatively low heart rate.

Shoulder and Back Loads

For cleaners involved in long periods of both cleaning tasks and kitchen tasks, the dominant shoulder was heavily loaded and we found no major difference between these tasks (see Figure 3). Therefore, it is unlikely that shoulder loads would be reduced through simply alternating cleaners between cleaning tasks and kitchen tasks. Rather, in order to introduce real variation into the load of the shoulder and the shoulder muscles, other work tasks should be considered. In contrast to our findings for the shoulder load, however, there does seem to be an appreciably higher load on the back when cleaning tasks are compared with kitchen tasks, since the time with heavy back load was longest during cleaning tasks (see Figure 4). However, the results indicate that, even during kitchen tasks, the back was nevertheless still heavily loaded about one-fifth of the time. Substituting cleaning tasks with kitchen tasks, then, may be used to reduce the load on the back to a certain extent but this clearly will not eliminate the back load. As in the case of variations in heart beat, part of the similarity in impacts upon the back and shoulders may be explained by the very small difference between the tasks examined—such tasks are both primarily performed either standing or walking. Consequently, interspersing these activities with tasks involving sitting could potentially provide sufficient physical variation to reduce back and shoulder strain. Again,

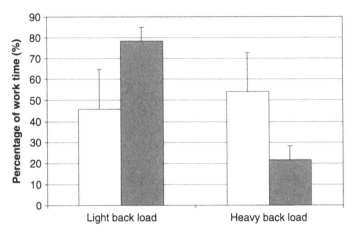

Figure 4: Mean (SD) duration of back load (% of time) is illustrated for cleaning (light bars, *n* = 9) and kitchen work (dark bars, *n* = 4).

such work tasks could perhaps be the planning task in a self-organized team.

Psychosocial Aspects

With regard to the psychosocial aspects of the study, some important differences were found between the cleaners from the three hospitals (see Table 4). Hence, the cleaners at hospitals A and B reported significantly poorer cognitive demands compared to those at hospital C, where such demands were at a level roughly equivalent to the national average (Kristensen 2001). The same difference was found for the possibilities for development. This means that cleaners in hospital C experienced much better possibilities for development than did those in either hospital A or B, although at all three hospitals these possibilties were generally far below the national average for Danish employees (Kristensen 2001). Finally, at hospital C cleaners expressed a significantly higher sense of coherence compared to those

Table 4: The scores of cognitive demands, possibilities for development illustrated for the three hospitals. The values are average index values related to a scale ranging from 0 to 100. Values above 60 and below 40 are considered statistically different from the Danish national average (Kristensen 2001)

	Cognitive demands	Possibilities for development
Hospital A	7	13
Hospital B	8	7
Hospital C	52	33

in hospitals A and B. Given the generally poor psychosocial environment among these cleaners, all-in-all we would have expected them
to have reported a low work ability index and sense of coherence.
However, the fact that they reported a work ability index in the best
half of the index and a sense of coherence index in the middle of the
index may be related to the fact that hospital C provides a greater
variety of work tasks for its cleaners (portering tasks at hospital C,
for instance, are conducted in combination with cleaners' organization
into self-operating teams).

Conclusion

The results of our examination of the ergonomic aspects of cleaners'
work show that repetitive muscular work of the arms associated with
high levels of static and dynamic muscular strain are quite common in
the most time-consuming cleaning tasks. Significantly, the muscular
load of the shoulders was high for both older and for more recently
developed floor-cleaning equipment, as well as for the cleaning of
surfaces above floor level. Moreover, the levels of strain were of such
an order as to represent a high risk of cleaners developing musculoskeletal disorders. Our results show that simply having cleaners switch
between different ways of doing the same cleaning tasks—mopping
back and forth versus using a figure of eight motion—does not seem
to be sufficient to reduce the overall physical workload. Rather, the
ergonomic analysis reported here demonstrates that it is necessary to
introduce variation into the cleaning work through job enlargement.
Put another way, in order to reduce the risks of short-and long-term
bodily damage, the employees engaged in professional cleaning must
be allowed to perform work tasks with a different strain profile on the
body.

In addition to such musculoskeletal and cardiovascular concerns
related to cleaners' work, we found the psychosocial environment
to be generally poor at the hospitals. Nevertheless, performance of
different types of work tasks and working in self-operating teams
seems to improve the psychosocial environment. In order to optimize
the physical load on cleaners and to provide a more mentally stimulating work environment, then, implementation of new work tasks,
which add a real variation to the load on the body, must be taken into
consideration when designing cleaners' workdays. Although the study
presented here only includes small groups of cleaners and should not
exclude further exploration of possibilities for job enlargement for
cleaners, its results nevertheless call attention to the fact that substituting cleaning tasks with other unskilled manual tasks such as kitchen
tasks and portering tasks only has a limited effect, and that a more
comprehensive strategy to design a healthier workday for cleaners is

necessary. Whether this is likely, given competitive pressures towards increased outsourcing and privatization, is, of course, the question.

Endnotes

[1] In 2002 ISS announced its intention to merge with one of its main domestic competitors, Sophus Berendsen. Through such a merger ISS, which focuses on cleaning and day-to-day maintenance of buildings and offices, and Sophus Berendsen, which often holds the contract for linen and laundry services within the buildings and offices serviced by ISS, would have created an even larger, vertically-integrated, yet sectorally-diverse, cleaning company. In the end, Berendsen was purchased by the Davis Service Group PLC, a British firm providing linen rentals, work wear rentals, uniforms for the health care, hospitality, cleanroom, manufacturing, and hazardous environments industries, and laundry services through subsidiaries Sunlight Service Group, Spring Grove Services, and Modeluxe Linge.

[2] One provider of equipment, for instance, claims that a new-style "Flat Mop Combo Bucket system" can reduce total cleaning time by 15% (see http://www.parish-supply.com/versamatic.htm; last accessed 9 November 2005).

[3] APDF measures the relationship between the amount of electrical activity and the force in a muscle. Thus the static level is the activity level that is exceeded during 90% of work time, the median level is the average activity, and the peak load is the level that is exceeded in 10% of work time. Although the peak level can be a risk factor, because high force demands can lead to mechanical damage of the tissue, in many modern work tasks the static work load may be the most important risk indicator because prolonged sustained activity, even at a relatively low force level, can metabolically exhaust the active muscle fibers.

[4] MVC is a measure of muscle strength. It can be expressed as either force (lbs, Newtons, etc) or as a moment around a joint (foot-lbs, Newton-metres, etc). A measurement of 10% MVC means that the muscle works at a force level corresponding to 10% of its maximal strength.

[5] MPF is a measure of how the power and frequency of the electrical signal to the muscle changes with fatigue, since it reflects the conductive velocity of the electrical signal and the degree of synchronization of the signals from the nervous system. RMS is a measurement of the intensity of the electrical signal.

[6] The *trapezius* muscle stabilizes the neck and the shoulder girdle and rotates and lifts the *scapula*; the *deltoid* muscle connects the *clavicle* with the upper part of the arm, allowing extension, flexion, and abduction of the arm; the *extensor carpi radialis brevis* and the *extensor carpi radialis longus* connect the *humerus* to the second and third metacarpals, respectively, so allowing the wrist to be extended and contracted. For a good set of illustrations of such muscles, see http://www.rad.washington.edu/atlas.

[7] Typically, wet mops consist of myriad lengths of strings of absorbent material which is submerged in water and/or cleaning solvent. Flat mops, on the other hand, typically have a rectangular head which swivels on the bottom of the handle. An important difference between these systems is that the wet mop and the mini mop both have to be carried during use so as to allow the long strings to move around. The flat mop, on the other hand, rests on the floor during use.

[8] This means that during 90% of the worktime the muscles in the resting shoulder showed constant activity at a level where it may present a risk of muscle fatigue.

[9] This means that the variation in heart rate was calculated as the difference between the low level that was exceeded during 90% of the workday and the peak level that was exceeded during only 10% of the workday.

[10] This test is based on a linear relationship between heart rate and oxygen consumption. During a bicycle test the workload can be controlled and therefore oxygen consumption can be estimated. This is significant because the lower the heart rate for a given workload/oxygen consumption, the better the aerobic power. In the one-point bicycle test, the workload is estimated from the self-reported normal activity level and daily exercise aiming towards a heart rate of approx 140 beats per minute.

[11] A Jamar dynamometer is a device commonly used in the medical field to measure grip strength.

[12] These determinations are based upon Danish government guidelines regarding the lifting of weight (see Danish Working Authority, "Departmental order regarding manual handling, No. 1164, December 16, 1992", and "Guidance on Interpretation, Lifting, Pushing and Pulling, September 31, 2005"). These guidelines have been described in Christensen, Pedersen and Sjøgaard (1995).

[13] For further description of these measures, see Jensen et al (2002) and Kristensen et al (2002).

[14] The concept of work ability as measured by the WAI can be defined as the ability of workers to perform their job, taking into account the specific work demands they face, their individual health conditions, and their mental resources (eg the ability to enjoy daily activity, feeling active and alert, and having hope for the future) (see Tuomi et al 1998). The index is calculated on the basis of six items: current work ability compared with lifetime best; work ability in relation to the demands of the job; estimated work impairment due to diseases; sick leave taken during the previous 12 months; and mental resources.

[15] The index consists of nine questions, which focus upon a worker's ability to clarify and structure causes of stress, to mobilize coping mechanisms, and to approach a problem situation as a challenge to be met rather than as an insurmountable obstacle (Setterlind and Larsson 1995).

References

Aguiar L L M (2001) Doing cleaning work "scientifically": The reorganization of work in the contract building cleaning industry. *Economic and Industrial Democracy* 22(2):239–269

Åstrand I (1960) Aerobic work capacity in men and women with special reference to age. *Acta Physiologica Scandinavica Supplementum* 1–92

Åstrand P-O and Ryhming I (1954) A nomogram for calculation of aerobic capacity (physical fitness) from pulse rate during submaximal work. *Journal of Applied Physiology* 7:218–221

Björkstén M, Itani T, Jonsson B and Yoshizawa M (1987) Evaluation of muscular load in shoulder and forearm muscles among medical secretaries during occupational typing and some nonoccupational activities. In B Jonsson (ed) *Biomechanics X-A* (pp 35–39). Champaign, IL: Human Kinetics Publishers

Bonde J P and Jensen L D (2004) Arbejdsmiljø og helbred hos rengøringsassistenter i Århus amt. En undersøgelse af sammenhænge mellem øvre bevægeapparatsymptomer og lidelser ved professionelt rengøringsarbejde (in Danish). Denmark: Kvindeligt Arbejderforbund and Århus University Hospital

Bonjer F H (1971) Energy expenditure. In L Parmeggiani (ed) *Encyclopedia of Occupational Health and Safety* (pp 458–460). Geneva: International Labour Organisation

Borg V and Burr H (1997) Danske lønmodtageres arbejdsmiljø og helbred 1990–95 (in Danish). Copenhagen: National Institute of Occupational Health

Burr H (1998) Ensidigt gentaget arbejde i udvalgte sektorer på det danske arbejdsmarked 1996–97 (in Danish). Copenhagen: National Institute of Occupational Health

Christensen H (1986) Muscle activity and fatigue in the shoulder muscles of assembly-plant employees. *Scandinavian Journal of Work, Environment and Health* 12:582–587

Christensen H, Finsen L, Hansen K and Jensen C (2000) Muscle strength in relation to type of work and musculoskeletal symptoms (pp 5775–5777). Abstract, International Ergonomics Association XIVth Triennial Congress and Human Factors and Ergonomics Society 44th Annual Meeting, 29 July–4 August 2000, San Diego, California

Christensen H, Pedersen M B and Sjøgaard G (1995) A national cross-sectional study in the Danish wood and furniture industry on working postures and manual material handling. *Ergonomics* 38:793–805

Davies C T M and Knibbs A V (1971) The training stimulus: The effects of intensity, duration and frequency of effort on maximum aerobic power output. *Internationale Zeitschrift für Angewandte Physiologie Einschliesslich Arbeitsphysiologie* 29:299–305

Davis K G, Marras W S and Waters T R (1998) Reduction of spinal loading through the use of handles. *Ergonomics* 41:1155–1168

Dempsey P G and Mathiassen S E (2006) On the evolution of task-based analysis of manual materials handling, and its applicability in contemporary ergonomics. *Applied Ergonomics* 37:33–43

Faber A, Hansen K and Christensen H (2006) Muscle strength and aerobic capacity in a representative sample of employees with and without repetitive monotonous work. *International Archives of Occupational and Environmental Health* 79:33–41

Hagner I-M and Hagberg M (1989) Evaluation of two floor-mopping work methods by measurement of load. *Ergonomics* 32:401–408

Hopsu L (1997) "Kehittävän työnohjauksen vaikutus siivoustyön kuormittavuuteen" (in Finnish, summary in English). Master's thesis, University of Helsinki

Jensen B R, Schibye B, Søgaard K, Simonsen E B and Sjøgaard G (1993) Shoulder muscle load and muscle fatigue among industrial sewing-machine operators. *European Journal of Applied Physiology* 67:467–475

Jensen C, Finsen L, Søgaard K and Christensen H (2002) Musculoskeletal symptoms and duration of computer and mouse use. *International Journal of Industrial Ergonomics* 30:265–275

Jonsson B (1978) Quantitative electromyographic evaluation of muscular load during work. *Scandinavian Journal of Rehabilitative Medicine Supplement* 6:69–74

Kristensen T S (2001) A new tool for assessing psychosocial work environment factors: The Copenhagen Psychosocial Questionnaire. *Arbete och Hälsa* 10:210–213

Kristensen T S, Borg V and Hannerz H (2002) Socioeconomic status and psychosocial work environment: Results from a Danish national study. *Scandinavian Journal of Public Health* (Suppl 59):41–48

Krüger D, Louhevaara V, Nielsen J and Schneider T (1997) *Risk Assessment and Preventive Strategies in Cleaning Work*. Bremerhaven: Wirtschaftsverlag NW

Kuorinka I, Jonsson B, Kilbom Å, Vinterberg H, Biering-Sørensen F, Andersson G and Jørgensen K (1987) Standardised Nordic questionnaires for the analysis of musculoskeletal symptoms. *Applied Ergonomics* 18:233–237

Laursen B, Søgaard K and Sjøgaard G (2003) Biomechanical model predicting electromyographic activity in three shoulder muscles from 3D kinematics and external forces during cleaning work. *Clinical Biomechanics* 18:287–295

Louhevaara V, Hopsu L and Sjøgaard K (2000) Cardiorespiratory strain during floor mopping with different methods (pp 5518–5520). Abstract, International

Ergonomics Association XIVth Triennial Congress and Human Factors and Ergonomics Society 44th Annual Meeting, 29 July–4 August 2000, San Diego, California

Nielsen J (1995) "Occupational health among cleaners" (in Danish, summary in English). Unpublished PhD dissertation, University of Copenhagen, and the National Institute of Occupational Health, Copenhagen, Denmark

Sanders S (2001) "Ergonomics" is Not a Dirty Word: New products are popping up to make cleaners' jobs easier with less strain and pain, but will distributors buy into the trend? *Sanitary Maintenance* September. Available at http://www.cleanlink.com/sm/article.asp?id=557 (last accessed 9 November 2005)

Schibye B and Christensen H (1997) The work load during waste collection and meat cutting among workers in different age groups. *Arbete och Hälsa* 29:272–278

Setterlind S and Larsson G (1995) The stress profile: A psychosocial approach to measuring stress. *Stress Medicine* 11:85–92

Sjögren B, Fredlund P, Lundberg I and Weiner J (2003) Ischemic heart disease in female cleaners. *International Journal of Occupational and Environmental Health* 9(2):134–137

Søgaard K (1994) "Biomechanics and motor control during repetitive work. A biomechanical and electromyographical study of floor cleaning." Unpublished PhD dissertation, University of Copenhagen, and the National Institute of Occupational Health, Copenhagen, Denmark

Søgaard K, Fallentin N and Nielsen J (1996) Work load during floor cleaning. The effect of cleaning methods and work technique. *European Journal of Applied Physiology* 73:73–81

Søgaard K, Laursen B, Jensen B R and Sjøgaard G (2001) Dynamic loads on the upper extremities during two different floor cleaning methods. *Clinical Biomechanics* 16:866–879

Tuomi K, Ilmarinen J, Jahkola A, Katajarinne L and Tulkki A (1998) *Work Ability Index*. 2nd revised edition. Helsinki: Finnish Institute of Occupational Health

United States Environmental Protection Agency (2002) Using microfiber mops in hospitals. Environmental Best Practice for Health Care Facilities, November. Available at http://www.ciwmb.ca.gov/wpie/healthcare/epamicromop.pdf (last accessed 10 November 2005)

Winkel J, Ekblom B, Hagberg M and Jonsson B (1983) The working environment of cleaners. Evaluation of physical strain in mopping and swabbing as a basis for job redesign. In T O Kvålseth (ed) *Ergonomics of Workstation Design* (pp 35–44). London: Butterworths

Woods V, Buckle P and Haisman M (1999) *Musculoskeletal Health of Cleaners.* Sudbury, Suffolk: HSE Books

Section 3

Chapter 10

Introduction: Cleaners' Agency

Andrew Herod and Luis L M Aguiar

It has become accepted wisdom amongst neoliberals that unions are an irrelevancy in the post-industrial economy. Indeed, for Newt Gingrich, former Speaker of the US House of Representatives, architect of the 1994 Republican takeover of the US Congress, and apostle of Third Waver Alvin Toffler (1980), the era of high levels of unionism in the mid-twentieth century facilitated by central government pursuit of Keynesian macro-economic policy was an aberration of historical development. Thus, for him, whilst "the Industrial Revolution herded people into gigantic social institutions—big corporations, big unions, big government—the Information Revolution is breaking up these giants and leading us back to something that is—strangely enough—much more like Tocqueville's 1830s America" (Gingrich 1995:63). Such a representation of union membership decline over the past two to three decades—one wherein decline is portrayed as returning us to the natural order of things, to an idealized arcadian economic past in which rugged individualism reigned supreme and employers and their workers could interact with one another without the need for some third-party intermediary—has been a powerful and central aspect of neoliberal discourse. Certainly, with union density seemingly in free-fall in a number of industrial nations, to many such a Gingrichian explanation appears to have some purchase.[1]

However, as Marx (1981) famously noted, if there were no difference between appearance and reality, there would be no need for science. In light of this distinction, the essays in this section all show in various ways how the ebbs and flows of unionism are being shaped not by some naturalistic economic rhythm but, rather, by political and economic practices such as outsourcing, privatization, deindustrialization, capital mobility, and a host of union-busting activities.[2] In return, the success of these practices in reducing levels of unionization is encouraging a significant degree of introspection as labor federations and individual unions in many of the nations of the global North debate, rethink, and redesign "best practice" models for organizing and for making themselves relevant to workforces that are much more culturally diverse than they were during the early post-war period when unions enjoyed their highest levels of membership (Davis 1986; Heron 1996).

If levels of unionization are generally low in most industrial nations today, they are particularly low in the service sector. For their part, cleaners have historically been fairly "invisible" to the labor movement, both because of the nature of their work (often conducted out of sight) and, frankly, because some unionists have often seen cleaners as representing insufficient numbers (read: not providing enough dues monies) in individual workplaces to warrant mounting expensive campaigns to unionize them. But, as labor market deregulation has encouraged outsourcing and a growth in the number of cleaning companies, cleaning has become one of the fastest growing occupations in North America, Australasia, and Western Europe. Faced with an expanding industry, then, labor movements are being forced to intensify their efforts to organize cleaners. Arguably, this has been nowhere more evident than with the Service Employees' International Union (SEIU) "Justice for Janitors" (JfJ) campaigns which have been unmatched in recent years in bringing workers into unions and which, in December 2005, organized 5300 cleaners in Houston, Texas, in one of the most successful unionization drives of low-wage workers in any industrial nation in the past two decades (SEIU 2005).[3] Moreover, not only has the JfJ had some important successes in the US but, increasingly, it is also looked upon as a model for organizing cleaners globally, with its version of "social unionism"—combining community involvement, public demonstrations to shame employers into recognizing the union, and direct action methods—being highly regarded, even inspirational. This does not mean, of course, that the SEIU and the JfJ are beyond criticism or controversy (Milkman and Voss 2005). Still, it is difficult to identify a more successful approach to organizing cleaners anywhere.

Within this context, the three papers in this section deal in different ways with the challenges of organizing cleaners in the face of employer aggression and state assault on workers' rights. Whereas the JfJ campaigns may have caught the public's imagination in recent years, long before them there were the "night cleaners' campaigns" in 1970s Britain, when young socialist feminists put theory into practice by joining forces with cleaners to organize the industry. In this regard, in the first paper in this section, Sheila Rowbotham, a founder of the women's liberation movement in Britain and a participant in those early campaigns, provides a rich and insightful personal account of the twists and turns of attempting to organize cleaners, describing how women cleaners moved from being "invisible" workers to securing public awareness of their plight as a result of their gritty determination. The agency of cleaners is clear in Rowbotham's account, and it is an agency that remains evident in a new generation of cleaners who today seek to organize London's Canary Wharf, amongst other places.

The second paper, by Marcy Cohen, brings us back to the contemporary era wherein public sector workers, who until recently were thought to be relatively insulated from market forces, have been subjected to outsourcing, work intensification, privatization, and union busting by government. Often viewed as part of the public sector's "auxiliary" workforce, cleaners have been a particular target of this neoliberal assault. In her article, then, Cohen argues that the neoliberal policies of the present British Columbia government have had two principal goals with regard to the health care sector: introducing the market mechanism into health care work (including that done by "auxiliary" staff such as cleaners), and minimizing the possibilities of health care workers' representation by the fairly militant Hospital Employees' Union (HEU). The government has pursued this agenda by facilitating the privatization of the province's health service, by centralizing control in the health authorities in the lower mainland, and by encouraging "sweetheart deals" between cleaning companies and an acquiescent Local of the Industrial, Wood and Allied Workers of Canada. In response, the HEU has organized public protests, argued cases before the provincial labor board, reinvigorated its organizing, and engaged in strikes which have received wide public support. Though the new neoliberal regime has been unkind to the HEU, Cohen believes the union has emerged stronger from these struggles as it has had to develop new strategies to secure members amongst an increasingly culturally diverse workforce.

If there is one union that seems to have had consistent success in organizing a culturally diverse cleaning workforce, it is the SEIU. However, whilst many unions are seeking to emulate its JfJ strategy, the SEIU has increasingly been subject to scrutiny from some labor quarters for its alleged power centralizing tendencies. Thus, in the final paper in this collection, Lydia Savage explores some of the tensions that have arisen within the SEIU as it has implemented its JfJ campaigns. Specifically, she examines how the JfJ's success has spawned conflicts over where power should lie within the SEIU and the broader labor movement. Hence, whereas the union's national leadership has argued that the house of labor needs to restructure and merge into a small number of very large unions capable of going head-to-head with global corporations—a position that in 2005 resulted in the SEIU leading several of the largest US unions to break away from the American Federation of Labor-Congress of Industrial Organizations to form a new labor grouping—others have raised concerns that the SEIU's plan will result in a centralized, top-down, and potentially bureaucratic and undemocratic model of union structure. As a way to consider what the JfJ campaign means for the shifting terrain of power between the national leadership of the SEIU and local unions and their memberships, then, Savage investigates how SEIU Local 399 in Los Angeles has been impacted by the attempts of the national SEIU to "up-scale" its JfJ campaign through developing a statewide cleaners' union Local so as to better position itself in California to win organizing drives against, and contracts from, global cleaning companies. For Savage, the key sets of questions here concern how unions might negotiate competing desires to encourage the centralization and decentralization of power in the face of locally focused, but globally organized, capital and what this means for the likelihood of their success.

Endnotes

[1] In 2004 only 12.5% (7.9% in the private sector) of US workers were members of labor unions (US Census Bureau 2005), compared with over one-third of workers in the early post-World War II period. In Canada only 30.4% (fewer than 20% in the private sector; Jackson 2005:68) of workers belong to unions, a decline from 37.2% in 1984 (Jackson 2005:62, Table 2). In Australia rates of unionization have fallen from an overall 42% in 1988 to 23% in 2003 (46.9% in the public sector, and only 17.7% in the private sector) (Australian Bureau of Statistics 2004).

[2] Bronfenbrenner (2000) reports that fully half of all US private sector employers faced with a unionization drive in the late 1990s threatened to relocate their operations if workers unionized (the figure rose to 68% in more mobile sectors such as light manufacturing, warehousing/distribution, and communications). Additionally, 92% of employers forced workers to attend meetings to hear anti-union arguments, 75% hired management consultants to run anti-union campaigns, 75% distributed anti-union literature in the workplace, 70% mailed anti-union letters to their employees, 78% held one-on-one meetings between supervisors and their employees, and 25% actually discharged union activists. Such activities have greatly undermined union election win rates—whereas 51% of organizing drives were successful where there was no threat to relocate, only 32% were successful in workplaces where the threat was made.

[3] The only other union that comes close to this level of successfully organizing service workers is the Hotel Employees and Restaurant Employees (HERE), especially its Las Vegas Local (Rothman 2002).

References

Australian Bureau of Statistics (2004) Trade union membership. *Australian Labour Market Statistics* (April). Canberra: Commonwealth of Australia

Bronfenbrenner K (2000) Uneasy terrain: The impact of capital mobility on workers, wages, and union organizing. Report submitted to the US Trade Deficit Review Commission. http://www.govinfo.library.unt.edu/tdrc/research/bronfenbrenner.pdf (last accessed 1 January 2006)

Davis M (1986) *Prisoners of the American Dream*. London: Verso

Gingrich N (1995) *To Renew America*. New York: HarperCollins

Heron C (1996) *The Canadian Labour Movement*. Toronto: Lorimer

Jackson A (2005) Rowing against the tide: The struggle to raise union density in a hostile environment. In P Kumar and C Schenk C (eds) *Paths to Union Renewal* (pp 61–78). Peterborough, Ontario: Broadview Press

Marx K (1981) *Capital: A Critique of Political Economy*, Vol 3. Harmondsworth, UK: Penguin

Milkman R and Voss K (2005) New unity for labor? *Labor: Studies in Working-Class History of the Americas* 2(1):15–25

Rothman H (2002) *Neon Metropolis: How Las Vegas Started the Twenty-First Century*. New York: Routledge

SEIU (Service Employees' International Union) (2005) More than 5,300 Houston janitors choose to form a union with SEIU. http://www.seiu.org/property/janitors/campaigns/houston_victory_pg2.cfm (last accessed 3 January 2006)

Toffler A (1980) *The Third Wave*. London: Collins

US Census Bureau (2005) *Statistical Abstract of the United States*. Washington, DC: Department of Commerce

Chapter 11

Cleaners' Organizing in Britain from the 1970s: A Personal Account

Sheila Rowbotham

In March 1970, *Newsweek* announced the birth of a new feminist movement in Britain after a conference of 500 gathered in Oxford (Anon in *Newsweek* 1970a:49). Many of the women who travelled to Oxford had been radicalized by the movements of the previous decade: the Campaign for Nuclear Disarmament, Anti-Apartheid, opposition to the war in Vietnam, the radical student movement, and the American Civil Rights movement. Determined to raise our grievances as women, and inspired by the emergence of a Women's Liberation Movement in the US, we were also concerned about injustice and inequality in general. Our rebellion coincided with an upsurge of trade union militancy in Britain. A strike of Ford's sewing machinists for equal pay in 1968 signalled a new spirit among working women and, as a result, in 1970 the Labour MP Barbara Castle introduced her bill for equal pay. This bill was planned to come into effect by 1975.

Shortly after the Equal Pay Act was passed, the Tories came into power, led by Edward Heath. That August, the Conservative paper, *The Daily Telegraph*, reported that the government was worried about "the continuing wages 'explosion'" (Hughes in *The Daily Telegraph* 1970:1), whilst at the Conservative Party conference in October 1970 the new Prime Minister Heath declared a crackdown on welfare "scroungers" (Anon in *The Evening News* 1970b:13). To a confident generation of trade unionists determined to improve working class living conditions, this was akin to a declaration of war. The response was clear and angry. The 1970s were to be a period of turbulent industrial unrest in which thousands of people became drawn into militant activity (Kelly 1988:104–114). Though this is all well documented, it is less known that this decade was also a period of hope for low-paid workers, many of whom were women and immigrants. Significantly, the composition of the work force, and to some extent the trade union movement, was imperceptibly beginning to change in low-paid manufacturing jobs and in the public sector, and this would have an effect upon the decade's events.

Strikes by women workers combined with Barbara Castle's Equal Pay Act to highlight the issue of women's low pay. Not only did it quickly become evident that employers were ingeniously getting around the law on equal pay by regrading jobs and ensuring that the things women did were not marked up as "skilled", but the Act simply did not apply to many women whose work was not regarded as comparable to men's (Hanna in *The Sunday Times* 1971:65). In 1971 women's average earnings were £12 a week. This was less than 60% of the male rate for a 40-hour week (Bruegel in *Socialist Worker* 1971:7). However, there were swathes of women workers who actually earned less, including the invisible night cleaners who moved into the streets of the big cities after dark. Part of a growing host of casual workers who were outside the regulated economy and the trade unions, such women suffered unsocial hours and bad working conditions, earning around £9 or £10 a week, and considered themselves lucky if they had one week's paid holiday a year.

The Cleaners' Action Group

During 1970 and 1971, the Women's Liberation Movement mushroomed. In March 1971, 5000 people marched for "women's liberation" through the London sleet and snow. Among them was a night cleaner, May Hobbs, who carried a placard that read "The Cleaners' Action Group" (Bruegel in *Socialist Worker* 1971:7). May Hobbs was a fighter. Indignant at the conditions she knew as a night cleaner, she had made contact with members of the International Socialists (IS) (now the Socialist Workers' Party), a Trotskyist group. Friends of mine in IS had asked me to put a note round in the Women's Liberation Workshop Newsletter and, in the autumn of 1970, a crowd of women and one man packed into my bedroom in Hackney, East London, to hear May Hobbs tell us about her efforts to organize cleaners. The Night Cleaners' Campaign had begun.

Every Tuesday night at around 10pm I headed off into London's financial area with my friend and co-leafleter Liz Waugh. We prowled the streets on the look out for cleaners. They were not hard to detect among the few city workers left in the area; tired, walking heavily, and carrying plastic bags. "Excuse me", we said. "Would you like to join a union?" Then we would produce the blue and yellow printed leaflets from the Transport and General Workers' Union (T&G). We were met with blank looks, especially from the Afro-Caribbean women, many of whom, it dawned on us, had not come across unions before. I started to supplement the T&G material with hand-written efforts produced on a duplicator (ancestor of the photocopier) which was at the school where I worked part-time, asking: "Why do night cleaners get less than day cleaners? Do night work for such low pay? Why don't cleaners get full cover money? Work on understaffed buildings? Get no Sunday bonus? Often no holiday pay? Have no security? Can be sacked without notice?" (Cleaners' Action Group 1971, unpublished leaflet). The answer at the bottom of the page to these questions was, of course, because they were not unionized. Most women regarded us prestidigitators as we stuffed our T&G leaflets into their hands. A few, however, did join.

Inspired by May Hobbs and evangelical in our desire to improve things, we had tumbled into the Cleaners' Action Group with little understanding of the conditions the women were forced to accept so that they might work in this bargain basement of capitalism. We learned by meeting them on the streets, but most of all through the friendships we formed in the course of the campaign. The cleaners taught us how the gloss of the swinging sixties concealed a grim, subterranean poverty in British society. In 1970 the Child Poverty Action Group revealed that three million children were living in poverty in Britain (Jackson in *The Times* 1970:7). The women who went into cleaning were likely to be their mothers. Some were from the unskilled working class which had known poverty for generations and others were Irish, Afro-Caribbean, Asian, Greek or Spanish immigrants. Though a few were in their twenties and a few over 60, most were in their thirties and forties with husbands who were low-waged workers, ill or disabled. The women looked older than their age, for they hardly slept at all, snatching a few hours after the children went to school. The accumulative exhaustion was etched on their faces. They had no time for the meetings and demonstrations which for we young activists in Women's Liberation had become a way of life. Nonetheless, a few of them came on that first march in 1971 to hear May Hobbs call for "the self-organization of women at their workplace". A *Socialist Worker* report, written by Irene Bruegel, records how May emphasized the need to fight employers and "press for greater democracy within their unions" (Bruegel in *Socialist Worker* 1971:7).

We leafleters soon found this was a tall order for the women we were trying to recruit. Even joining a union was a major step. Many were too afraid because they were claiming Social Security, had immigration problems or were simply terrified of the contractors. Sally Alexander described in an account for the socialist feminist magazine *Red Rag* how at first we had simply imagined we would leaflet all buildings in London. Then we tried concentrating on one contractor and, when this proved difficult, focused on big buildings (Alexander 1994:259–260). Sally began to leaflet two enormous Shell buildings in West London where a cleaner, Jean Mormont, emerged as the shop steward. From a large family of 18 and the mother of seven herself, she remembered being in the Auxiliary Territorial Service (ATS) in the war as a kind of holiday (McCrindle and Rowbotham 1979:42). Despite her demanding life, she became one of the most steadfast supporters of the campaign. The women at Shell complained not only about their pay but about the inadequate staffing which forced them to cover ever more offices, working without proper equipment and in stifling air due to the air conditioning being turned off (Alexander 1994:260). We would hear similar objections from other women. In East London, Liz Waugh and I were on a fast learning curve about the contract system. We would laboriously unionize a building and come back to find the women scattered by the cleaning agency. Slowly we began to piece together a picture of the industry, partly from Jean Wright, who had been a cleaner for many years. She was solely responsible for a medium-sized block in the City and her teenage son and husband used to come in to help, assisted on leafleting nights by Liz and myself. As we all cleaned, Jean Wright would talk about the bizarre informal hierarchies in the business and explain how a good supervisor on a big building needed real planning skills. In the rackety cleaning business, however, merit did not always decide who was made a supervisor. Each firm operated in differing and apparently random ways.

The contracting of labour had been common during the 19th century in agriculture, the building trades and in government services. From the late 19th century reformers had campaigned against the system and pushed for direct, regulated employment, including equal pay for "char-ladies" in government buildings (Paul 1986:11; Rowbotham 1999:133–134). However, in the 1930s the growth of large offices had led to the first modern cleaning firms being formed. In the post-war era these had expanded and had received a recent boost in 1968, when the Labour government, keen to show they were making Civil Service cuts, had sacked 4000 directly employed cleaners (Alexander 1994:263). By the 1970s a few cleaning contractors had become big companies, but new firms were constantly appearing because it required very little capital investment to start up. The main cost was labour. The businesses on these lower rungs were often unstable. Indeed, we found some were extensions of criminal gangs who used overt intimidation.

The contract cleaning industry appeared marginal to the trade union movement in the early 1970s. Though the T&G had its roots in the late 19th century unionization of the unskilled and unorganized, those days were long gone. They were a big bureaucratic outfit and, though the leader Jack Jones was on the left of the Labour Party, the union officials saw recruiting cleaners as a waste of resources. The

Cleaners' Action Group was on the outside looking in when it came to the world of trade unionism. Neither May Hobbs, Jean Mormont nor Jean Wright had experience in negotiating trade union structures. Sally Alexander had been an actress and in Equity, and I was in the National Union of Teachers, but the byzantine rules of the T&G were double Dutch to us. Liz Waugh's mother Lucy, an East London working class woman who got involved in women's liberation along with her daughter, was equally perplexed. On being told the cleaners had to be in the window cleaners' branch she spent ages looking for their elusive branch meetings.

Exasperated by confusion and muddle, May Hobbs began to insist that the cleaners should have their own branch. The trouble was that we did not have enough women signed up for that, even though we found some cleaners were mysteriously already in the T&G. By the summer of 1971 we were at an impasse with the union. Some of the Women's Liberation Workshop leafleters were attracted by a proposal to create a women's union on the lines of the old Women's Trade Union League. Others of us argued the cleaners were too vulnerable as it was. An alternative idea was for a cleaning co-operative, but this was rejected because it would have meant setting ourselves up in business (Alexander 1994:259–260). Leafleters began drifting away, including the International Socialist women who went looking for the revolution elsewhere. By the autumn only a handful of us were left to produce an issue of the Women's Liberation Workshop magazine *Shrew* on the night cleaners. Liz and Lucy Waugh, Sally Alexander and the artist Mary Kelly, whose work with the night cleaners inspired her art work, took up this task. Mary was helping a left film group, the Berwick Street Film Collective (later called Lusia), to make a film of the campaign.

The truth was that the grand sounding "Night Cleaners' Campaign" was somewhat overblown and we were rather better known on the left than our actual numbers warranted. Our Night Cleaners' *Shrew* carried a report of a speech by the Irish socialist Bernadette Devlin (now McAliskey), elected MP in the wake of the Civil Rights movement in Ireland (Anon in *Shrew* 1971:6). She sat with her legs dangling from a table and addressed her rather scanty audience of night cleaners with her customary eloquence and passion. Assembling even these cleaners had been a Herculean task. May Hobbs's husband Chris brought some in his ancient car. Others were perched in relays on the back of Liz Waugh's somewhat alarming motor bike. Our campaign might be strong on speakers and writers, but it was weak on foot soldiers.

After *Shrew* came out, more leafleters appeared, including a woman from the International Marxist Group, another trotskyist group. She would quickly produce a pamphlet ("The Nightcleaners'

Campaign, c. 1972") through the "Socialist Woman" group linked to
her organization, saying what should be done; an act which we
fiercely democratic Women's Liberation types resented as high
handed. However, any problems of internal democracy were nothing
in relation to the continuing problems of engaging with the T&G.
May Hobbs, who was a natural-born direct actionist, had taken to
ringing Jack Jones at home and complaining to his wife. She was also
on good terms with the liberal media who happily reported the plight
of exploited night cleaners ignored by the union bureaucracy. This
did not endear us to the union. Even when we did manage to get
the T&G to meet with the cleaners, the communication gap was
marked. The film-makers from Lusia, Marc Karlin, Humphrey
Trevelyan and James Scott, caught a revealing moment at a meeting
in a pub with Jean Mormont, the Shell women and Mary Kelly. As
the official drones away in that peculiar language used by trade
unionists which is so impenetrable to outsiders, Jean Mormont,
black rings of tiredness beneath her eyes, slowly drifts away into the
music of the juke box in the background.

Cleaners' Strikes

May Hobbs decided we should focus on the Civil Service Union
(CSU), which had some cleaners as members. One obvious advantage
was that the CSU could draw on the support of their members inside
government buildings. Moreover, this was a markedly different style
of trade unionism. The union official, who was young and zippy and
drove a red sports car, was willing to come down to buildings and talk
to the cleaners during their break at 1am, a prospect which had always
freaked out the T&G men, who were nine-to-fivers. The CSU journal,
called *The Whip*, gave the campaign publicity. Our morale soared. In
the summer of 1972 cleaners on two Ministry of Defence buildings—
the Empress State and the Old Admiralty—went on strike (Anon
in *The Whip* 1972:1). They were in high spirits on the picket line,
partly because the CSU strike pay was £10 a week, which was more or
less what the women could earn whilst cleaning. The International
Socialists and the International Marxist Group came back to help the
leafleters from the Women's Liberation Workshop and the picket
grew. I can still remember a nagging guilt if I missed a day's picketing
to work on the book I was writing, *Hidden from History*.

These were militant times and the striking cleaners received instant
trade union support. The T&G lorry drivers refused to cross our picket
lines and supplies began to dry up in the Ministry of Defence, most
crucially the beer for the bar. Inside information from sympathizers
in the Empress State Building was that lack of beer was having a
terrible effect on morale. Post Office workers refused to deliver mail;

printers, railway workers and clothing workers sent donations. The local Trades Council came along with good practical advice about whom to contact in the area. One odd encounter was with some men at the Admiralty building one night who insisted we had to let them in because they looked after the tunnels. The tunnels, they explained, had to be kept in good order because the Queen and other important people would escape down them in the event of a nuclear attack. The Cleaners' Action Group was clearly threatening the very defence of the realm!

At the Empress State building in Fulham, the picket began to assume a carnival atmosphere. A nearby Italian restaurant allowed Lusia Films to use their electricity. The film makers rigged up a screen and began to show films, most notably *Salt of the Earth*, Herbert J. Biberman's wonderful 1953 film of a strike in a New Mexico mining community in which the women played a key role. Passionate, sensitive, humourous, *Salt of the Earth* resulted in him being black-listed during the McCarthy era, whilst the Mexican actress, Rosaura Revueltas, was repatriated to Mexico (Pym 2000:959–960). The cleaners, several of whom were from the Caribbean and Ireland, loved this drama in which class, race and gender interacted in ways that related closely with their own experience. Lusia Films had been inspired by the activist film making of the May events in Paris during 1968 and by early Russian revolutionary films. They were part of a creative new wave of documentary film makers who were just beginning to take off in Britain at that time. They raised money by doing advertisements and showed their films at meetings. Whilst some took a straightforward newsreel style, Lusia was experimenting with new forms of communicating (Dickinson 1999:126–136; Rowbotham and Beynon 2001:143–158).

Cleaners and feminists picketing, singing and dancing at the Ministry of Defence made a good story and the strike was covered widely in the media. Our targeting of high-profile government buildings brought results. The CSU was able to get the contractors to recognize the union. The strikers obtained a raise of £2.50 per week and a 50 pence night allowance. The women were joyous and at Empress State remained so confident that they were able to push their wages up to £21 a week, well above the average women's wage of £12 (Anon in *The Whip* 1972:1; Alexander 1994:262). The CSU, however, clearly found negotiating with contractors a nightmare and realized that all their efforts would be invalidated when the time came for the contract to be renewed. Whilst everyone else in the campaign rejoiced at the cleaners' victory, the Assistant General Secretary of the CSU, Les Moody, was mopping his brow. He told Martin Walker from *The Guardian*: "It's a labour of love. In time and effort it costs us a lot more than the membership fees we get

from them" (Walker in *The Guardian* 1972:63). Regardless of this
begrudging comment, a radical wing in the union was delighted.
The CSU began to press for the cleaning of government buildings to
be taken back in-house. The contractors, meanwhile, spoke gloomily
of the dangers of bankruptcy.

Loss of Impetus

Victory extended the fame of the Cleaners' Action Group. May
Hobbs, who had a gut understanding of spin long before the Blairites
discovered it, was increasingly away speaking around the country,
explaining how cleaners were flocking to join the union. No one knew
precisely how many cleaners there were because women were working without cards; the numbers in the union were equally confusing
because membership fluctuated. However, we were certainly not
recruiting these supposed hordes of cleaners. The reality of the
leafleters on the ground was far more mundane. In the summer of
1972, Liz Waugh and I started to recruit a group of four women into
the CSU on a building May and her husband Chris Hobbs had decided
we should target. It was Companies House at 207 Old Street, where
the records of registered companies were kept. My notebook recording
the receipt of dues describes them being paid £14 for a five-day week.
Their hours were 10pm to 6am, with one week's holiday pay. I went
there every week collecting between 5 p and 24 p a week from the
women until just before Christmas, when it was discovered that they
could not be recruited into the CSU after all. Hanging my head
in shame, I refunded the dues from my own pocket, feeling like a
fraudster. To my amazement the women treated me like a heroine.
They might be unfamiliar with the purpose of trade unions, but they
knew all about informal savings systems. It was customary for people
to pay for their turkeys at the butcher slowly over time; the small sums
I handed over to them seemed like Christmas bonuses—the turkey
money coming home to roost.

The leafleting stopped during 1973. This was partly because of our
exhaustion, but also because of internal tensions within the campaign.
Not only was there the yawning class gulf between the leafleters
and the cleaners but there was anger and unease among the women
cleaners themselves. Several women who had become involved
distrusted May Hobbs's leadership, and this was made worse as she
became, understandably, interested in other causes. She and her
husband Chris were great stirrers and rousers, but they were not
meticulous about details or good at building up a core of people to
work together. Jean Mormont and Jean Wright could do this locally,
but would defer to May Hobbs in relation to the Cleaners' Action
Group. During the strike at the Empress State building two women,

one Irish and one from the Caribbean, developed into an organic leadership. But we were never able to foster this process in the Cleaners' Action Group as an organization. The working class women in the group who had no previous experience of working in any organizational structures found it difficult to operate in a context which was not a purely personal network of women. Our ideology of sisterhood did not wash with the cleaners, whose relations with other women were complex and often conflictual—though, interestingly, these conflicts were not articulated in terms of race and ethnicity but, rather, in personal grievances that cut across these differing identities. Equally, because they were used to male leadership in daily life, these women were probably *more* suspicious of May Hobbs as a leader than they would have been of a man. We leafleters in Women's Liberation, however, were keen not to impose decisions on working class women. Those of us who had leafleted for a while had learned from the experience but we had a libertarian politics that deplored any inequality of knowledge. Consequently, we kept being steamrollered by women in the left groups who had no such reservations.

We blamed ourselves for failing the cleaners, though we were dimly aware that the contract system presented serious problems for unionizing. We had, of course, no idea that this form of work was going to be extended by a Conservatism that made Heath look benign. It was inconceivable that contracting out services could become the prevailing pattern for whole chunks of the British economy. But this was, of course, what happened in the 1980s when reducing regulated labour conditions by any means came to be seen as legitimate by the Thatcher government. More and more, vulnerable workers, including many women, were employed through the contract system and some of the big players in the industry transmogrified into service multinationals. Workers who had regarded themselves as the backbone of the labour movement found themselves in the company of women they had considered to be marginal. The shock was palpable and a generation of trade union militants never recovered.

Impact

Ironically, by the time Lusia films finished their long, experimental documentary *Night Cleaners* in 1975, there was no campaign operating but there was a great deal of interest in night cleaners, owing to May Hobbs's speeches to meetings and rallies, along with our middle-class knack for publicity. But when the film was shown at meetings, it provoked extreme reactions. Left audiences were used to the format of TV newsreels and were bewildered by Marc Karlin's efforts to create space for viewers to think, imagine, probe and question with blank screens and long, slow shots of the women's faces. In

refusing cinematic conventions he wanted to get beyond the exter-
nalities of "struggle" into the lives and feelings of the women. He
took some people with him, including some of the cleaners, but he left
others furious, including an irate May Hobbs, who had always wanted
a quick, short, propaganda film. Seen in retrospect, *Night Cleaners*
provides fascinating footage of the mass demonstrations against
Heath's policies, with one magical moment in which two young
miners dance together. It also chronicles a group who were rarely
portrayed with sympathy, the 1970s London poor, living on the edge,
the strata the Tories called "scroungers". It documents the people
who, by and large, go undocumented through history. Romantic and
conceptual at the same time, it explores the unseen; the city at night,
the invisibility of women's labour and the exhaustion permeating their
lives outside work. It was indeed about the hidden injuries of class
(Dickinson 1999:149–152; Rowbotham and Beynon 2001:152–153).
Night Cleaners became a classic work, recognized by film makers as
pioneering and stored in the British Film Institute archive. However,
the night cleaners, still largely ununionized, continued to go to work at
10pm each night carrying their plastic bags of belongings, though
cameras and leafleters no longer pursued them through the deserted
streets.

There were some spasmodic attempts to organize cleaners in other
places over the course of the 1970s. In Oxford, during the early 1970s,
the Women's Action Group, whilst leafleting the working class
housing estate of Blackbird Leys about nursery provision, made
contact with a group of women cleaners at the Cowley car plant.
Hilary Wainwright, who was in the Women's Liberation Movement
and the International Marxist Group, told them about May Hobbs and
contact was made with the local T&G. However, the T&G would not
allow the Women's Action Group or May Hobbs into their meeting
with the cleaners and so they stood angrily outside. Nevertheless, the
Oxford cleaners did become unionized and won some improvements
in wages and conditions.[1] Several attempts were made to organize on
college campuses. At Durham University, inspired by May Hobbs's
account of the "successful struggle to unionize London cleaners", a
student, Lynda Finn, and Gavin Williams, a lecturer, decided to try
to organize the college cleaners in 1973. It proved far more difficult
than they had envisaged owing to resistance from the University and
inter-union disputes (Finn and Williams c 1976:5).

Whilst organizing cleaners presented enormous problems, the publicity generated did contribute to a shift in attitudes in the labour
movement towards low-paid women workers, including cleaners.
During the 1970s feminists were extremely active in trade unions on
pay and conditions, as well as lobbying union branches, trades councils and the Trades Union Congress on social issues such as abortion

and nurseries. Women with expertise in the law and in the trade unions helped to link the two movements. The solicitor Tess Gill, together with an official in the white collar union AUEW-TASS (Linda Smith), began to explore how low-paid women workers could use existing legislation to strengthen their bargaining power (Anon in *Morning Star* 1976a). In contrast to the tendency which prevailed in the libertarian and trotskyist left to regard the state with suspicion, they represented an opposing tradition in the British labour movement of turning towards the state to counter capital. Socialist feminists were now insisting that women needed the state. A debate ensued within the women's movement on the welfare state which led some feminists to ask how to gain access to the resources controlled by the state in ways that helped the most vulnerable working class women (The London to Edinburgh Weekend Return Group 1980; Rowbotham, Segal and Wainwright 1979). Little did we know that this was the welfare state's eleventh hour.

Changing Circumstances

It was evident, however, that the economic context was undergoing a change. By the mid-1970s pressure from the IMF forced the Labour government to make cuts in public services. According to time-honoured practice these were directed at the most vulnerable. Initially the cuts were met by a confident and staunch resistance, which meant that the state itself increasingly became a site of conflict. In 1976, when the Area Health Authority in Birmingham decided they could no longer afford to employ extra staff, the cleaners at Mosley Hall Hospital refused to do more work and went on strike. Instant support came from male porters and hospital drivers in the South Birmingham Hospital District who refused to handle dirty linen from the hospital. Within 24 hours the Health Authority found that they could, after all, employ more cleaners (Anon in *Spare Rib* 1976b:21–22).

In the late 1970s, when low-paid workers rebelled against wage restraint, an extraordinarily powerful media myth took shape which conveniently happened to bolster the interests of both the Labour and Conservative Parties. According to the new script, workers like those Mosley Hall cleaners were portrayed as greedy and lazy. The rest, of course, is history; Thatcherism rode to power on the myth which has never been dislodged. After Margaret Thatcher was elected in 1979, not only did inequality increase in British society but it became inadmissible to argue for the redistribution of wealth. The Tory tactic of privatizing public services had not been part of the original plan; it was developed *ad hoc* after an experiment devised by a Conservative Councillor, Christopher Chope, in the South London borough of Wandsworth proved popular. In 1983 the council targeted the dust-binmen, a group of workers who were not loved by the public because of a long and smelly strike and a productivity scheme which meant that they often left debris in their wake as they rushed to empty the bins. Rubbish collection was privatized.

Privatization was accompanied by changes in legislation nationally which affected trade union action and the position of low-paid workers. In 1983, the 1946 Fair Wages Resolution, which required central government contracts to employ workers on wages and conditions which were not less favourable than those agreed by the unions in the trade or by the general level of pay in the type of work, was scrapped (Pearson 1985:85–99). Thatcher was not able to abolish the Wages Councils which fixed rates in low-paid industry but the next Tory Prime Minister, John Major, got rid of them. An all-party consensus that the state had an obligation to protect low-paid and vulnerable workers—a consensus which owed much to reformers, including feminists, in the late nineteenth and early twentieth centuries and which had crystallized after World War Two—was shattered by Thatcher and Major.

The U-turn in state policy, the adoption of privatization on a large scale and the collapse of manufacturing industry during the 1980s meant that cleaners who ten years before had seemed so peripheral in the labour movement started to come to the fore. A group in the private sector working for a large West End London store consciously developed the link between the community and the workplace which had arisen accidentally in Oxford. A workforce of Latin American immigrants, some of whom were highly educated and in flight from repressive regimes, built up confidence by dealing with individual grievances, helped by the North Kensington Law Centre. They then unionized successfully through the T&G, which was becoming more open in its approach. They produced a newsletter called *El Mopo* (*The Mop*) and were able to raise their wages (Pearson 1985:42–51).

It was, however, the public sector which saw the most intense contests. Cleaners who were employed in the public sector did not necessarily have higher wages than those who were contract workers. They were, however, more likely to be covered for sickness, holiday pay and pensions. There were several battles against privatization and Asian women workers, a new force in the British trade union movement, played a prominent part in these. Organized by the National Union of Public Employees, South Asian cleaners at Hillingdon Hospital in West London protested against privatization (Paul 1986:67). They were not successful. However, in Hackney, East London, in 1984, after all the health workers in the borough went on strike for one day, the Area Health Authority decided that it would be a bad idea to put domestic services out to competitive tender (Paul 1986:70). Despite this success, the problem remained that the unchanging, unflinching resolve of central government made it difficult to sustain resistance against privatization.

Women cleaners who were already working for contractors found that the introduction of competitive tendering resulted in a further decline in pay and conditions. Barking Hospital in northeast London saw a long and bitter dispute which arose in 1984 when the cleaning company Crothalls underbid the contract they had formerly held by cutting pay and holiday provision and putting cleaners on flexible shifts. The shifts played havoc with the women's lives and were particularly resented by those who were single mothers with children because they could not plan their time. A long-standing cause of exasperation on the part of the cleaners was their claim that they were given inadequate cleaning materials. By highlighting the negative impacts of scabs' lack of knowledge of the patients, in a roundabout sort of way the Barking cleaners who went on strike were able to show the importance in their own work of the tacit knowledge and skills learned through doing the job over time, caring skills which were not included in how their work was evaluated. Indeed, the negative consequences of the short-term policy of cost cutting on cleaning quickly became evident as the Environmental Health Officer's Report in April 1984, one month after the strike began, found the cleanliness of Barking Hospital to be unsatisfactory (Paul 1986:45–47).

During the 1980s, the combination of publicity generated by campaigners and strikers, along with the government's resolve to endorse the contract system, resulted in more research being done on cleaning. A comprehensive ACAS (Advisory, Conciliation and Arbitration Service) report in 1981 on contract cleaning recorded a deterioration in pay and conditions during the 1970s. A joint CSU/Low Pay Unit Report on cleaners who were directly employed showed that whilst their wage levels were similar to contract cleaners, their sickness, holiday and pension provision was better. In 1983 a useful report produced by the Incomes Data Services (IDS) noted how paid holidays in some large, private firms such as British Leyland and Ford for directly employed cleaners included in general wage negotiations were between 20 and 25 days, much longer than those of contract cleaners (IDS 1983:1–9). Radical community projects such as Community Action and Public Service Action began to document privatization nationally and to provide advice for campaigners. In 1984 the magazine *Community Action* recorded support for cleaners from Inland Revenue workers in Llanishen, Wales, and Addenbrooke's Hospital in Cambridge. It also described how the T&G had ensured that contract cleaners would be included in the general bargaining structures of a factory called Hickson and Welch Ltd in Castleford in the North of England (Anon in *Community Action* 1984:69).

Left-wing local authorities began to put their weight into researching labour conditions. The pioneer was the Greater London Council
(GLC), led by Ken Livingstone, which set up an Industry and
Employment Unit in 1983 to produce a participatory industrial strategy for London. I went to work for the Unit at the end of 1983 and in
1984, determined that the cleaners would not be forgotten, produced
the committee paper on cleaning which brought them into the strategy.
Rejecting the bland style customary in committee papers, I confused
the Tories with quotes from John Ruskin and Harold Macmillan on
the value of cleaners and the iniquities of mean-minded contractors.
Irene Bruegel and other women at the GLC went on to develop an
innovatory programme of reform in pay and benefits for the Council's
own cleaners, including the "Basic Skills Project". This allowed
the cleaners access to flexible education and training whilst at work.
Anything from illiteracy to vocational needs could be catered for.
Many were supporting families and keen to get out of the trap of
low-paid work. A few decided they wanted to go on to study further.
The time to think and discuss also led some women to speak up in
union meetings (Paul 1986:38).

The Industry and Employment Unit was able to compensate for the
lack of resources of low-paid London workers, including cleaners, in
many small ways. The GLC librarian disseminated information about
cleaning companies which was available to investors but was too expensive for cleaners to obtain. In 1985 I was able to commission a
report into cleaning by a Law Centre worker, Jane Paul. *Where There's
Muck There's Money* appeared in 1986, just before the GLC was
abolished by central government. Nonetheless, it circulated widely
in London, documenting the destructive implications of competitive
tendering in hospitals, schools and even on the London Underground.
It also showed how the contract system was an international phenomenon, citing two reports—"Government for Sale" (1977) and "Passing
the Buck" (1983)—concerning contracting out in the US written by
the former *Washington Post* journalist John D. Hanrahan and produced by the American Federation of State, County and Municipal
Employees (AFSME). Jane also drew on the work of the British medical sociologist Geoff Rayner and Gina Glover from the Wandsworth
Photo Co-op who had gone to investigate the North American
experience in 1983 and had been impressed by the fact that the public
sector unions in the US had the legal right to consultation and negotia-
tion when services were contracted out to private firms, which was
not the case in Britain. Significantly, they had observed that many
contracts in the US were in the hands of Crothalls, the company with
which the Barking cleaners had conflicted (Paul 1986:78–81).

The daunting fact that cleaning was a multi-national industry was
beginning to dawn. One of the trade union research groups funded by

the Industry and Employment Unit at the GLC began to develop international links with trade unions and cleaners groups. In 1987 an International Cleaners' Conference entitled "Invisible Workers" was held in London, bringing together cleaners from across Europe. It revealed how big multi-national companies were operating in Europe and how the labour force, too, was international. Among those attending were migrant workers from many lands, including North Africa, Latin America and Turkey. They voiced familiar complaints about the companies who employed them: of having to cover more rooms than was agreed upon; of inadequate, even unsafe, cleaning materials; of unhygienic conditions in hospitals; and, of course, of low pay (Gowen 1988:20–22). Though the trade unions were beginning to look towards their counterparts in Europe, well-organized skilled workers still looked glazed at the suggestion that cleaners might be included. The prospects for international links between workers at the bottom of the pile were not high on the agenda of an increasingly battered trade union movement.

Nonetheless, a memory of resistance survived with surprising tenacity. In the autumn of 1995, low-paid ancillary workers at Hillingdon Hospital in West London went on strike after their contract cleaning employers, Pall Mall, cut their already-low wages. Some of the strikers were the same women who had protested against privatization a decade before. The mainly South Asian workforce was driven by a deep sense of injustice and refused to stop picketing. They took their cases to the Industrial Tribunal and, eventually, after four years, won compensation.[2] During their long dispute they went around speaking in many countries, as well as in Britain, and close links developed between them and other workers, for by the mid-1990s the casualization of work had reached groups who had previously been part of the well-organized workforce. In 1995 the Hillingdon women marched alongside Liverpool dockers and their families striking against casualization, with the dockers donating money out of their own strike fund to the Hillingdon women. For Britain's cleaners, in the 1990s new solidarities came out of shared adversity. However, the dockers who were resisting the global grain were defeated and by the early twenty-first century casualization had spread steadily up the social hierarchy to reach professional workers. Originating in the fringes of the hidden economy, it had now come to characterize society as a whole.

A New Phase of Organizing?

The North American "Justice for Janitors" campaign and the 1998 Ken Loach film about it ("Bread and Roses") has recently stimulated new attempts to organize cleaners on the big Canary Wharf building in London's transformed docklands area. Researcher Hsiao-Hung Pai reported in *Feminist Review* (2004) that undocumented contract workers were being employed there without fixed holidays and sick pay entitlements. She was told by other workers that they had observed some undocumented migrant workers being dismissed without verbal or written warnings (Pai 2004:165–172). Working in combination with the East London Communities Organisation (TELCO), supporters have persuaded the cleaning company ISS, which is a leading multi-national firm, to recognize the union. In October 2004 Tania Branigan reported in *The Guardian* newspaper that the workers with legal contracts were earning £5.20 an hour and had 12 days of holiday time and 8 public holidays a year, though they had no sick pay or pensions. The T&G, however, considered £6.70 to be the minimum wage for workers in London and argued that they should receive sick pay, pensions and longer holidays. Mayor Ken Livingstone supported the union's case. Nevertheless, the T&G had to call off a demonstration scheduled to coincide with the European Social Forum in London after the owners of Canary Wharf had an injunction taken out on the grounds that there were no public access rights to the building (Branigan 2004:14). Despite this failing, Colin Cottell noted the focus on the financial companies at Canary Wharf has subsequently produced several successful wage increases and paid holidays, a result which indicates that companies are accepting some degree of responsibility for the contracted out cleaning jobs (Cottell 2005:12). The danger, however, is as it always has been that these modest gains might be offset by companies reducing the staff and thus intensifying the work load. Thus, as Ken Loach has observed: "This will be a long war with many battles . . . the buck stops with those who hand out the contracts" (*The Guardian* 2004:14).

Thirty years ago, the Cleaners' Action Group could never have foreseen that cleaners were going to become part of the global economy. Yet those huddles of women we leafleters approached in the London night were indicative of an exploitative and shortsighted system of employment which would be massively extended world-wide. In the 1980s and 1990s the Barking and Hillingdon women had tried to warn of the wider consequences of cheap labour and their voices went unheeded. By 2005, however, Helen Carter would report in *The Guardian* that 100,000 patients a year were getting hospital-acquired infections resulting in around 5000 deaths due, in part, to inadequate cleanliness. Unfortunately, this macabre reality has not led to a greater appreciation of skills and value of cleaners. It has, however, resulted in a technological innovation. The Airedale NHS Trust Hospitals in Yorkshire have introduced new, more efficient microfibre mops to fight the rise of the bacteriological "super bugs" (Carter 2004:10). Microfibre mops in a hi-tech age may seem a small advance, but change comes slowly in the cleaning labour process. Just how slowly would have been inconceivable to those of us who eagerly set up the Cleaners' Action Group in 1970. In an odd way, our ignorance and inexperience gave us the courage to fight against a system, the power of which we did not comprehend. In our naivety and outrage we stumbled upon something that was far, far bigger than anyone at the time envisaged.

Endnotes

[1] I am grateful to Hilary Wainwright for information on the Oxford Cleaners' Campaign.

[2] I am grateful to Francis Reynolds, former Hillingdon cleaner, for information on the dispute in the 1990s.

References

Alexander S (1994) *Becoming a Woman and Other Essays in 19th and 20th Century Feminist History.* London: Virago

Anon (1970a) Tory plan to crackdown on the scroungers. *Evening News* 8 October

Anon (1970b) Britain: The worm is not for turning. *Newsweek* 23 March

Anon (1971) Bernadette Devlin speaks to cleaners. *Night Cleaners'* Special Issue, Shrew, December

Anon (1972) Something to smile about . . . as the union sweeps to "double" victory for contract cleaners. *The Whip: The Civil Service Union Journal* September

Anon (1976a) How Elsie Bucket and Iris Moppit won 20p. *Morning Star* 4 October

Anon (1976b) Changing their tune. *Spare Rib* 45

Anon (1984) Cleaners issues. *Community Action* 67

Branigan T (2004) Loach pitches in for low-paid cleaners. *The Guardian* 9 October

Bruegel I (1971) Women marchers stress need for union action over equal pay. *Socialist Worker* 13 March

Carter H (2004) Hospital mop and bucket become history. *The Guardian* 26 November

Cottell C (2005) It's time rich city firms cleaned up their act. *The Guardian* 7 May

Dickinson M (ed) (1999) *Rogue Reels: Oppositional Film in Britain 1945–9.* London: British Film Institute

Finn L and Williams G (c 1976) *The Durham Experience: Bureaucrats and Women Cleaners.* London: A Solidarity Pamphlet

Gowen S (1988) Invisible workers: An international cleaners' conference. *Spare Rib* January

Hanna V (1971) Equal pay: Industrial apartheid as firms evade the Act. *The Sunday Times* 21 November

Hughes A (1970) Pay demands worry government. *The Daily Telegraph* 3 August

IDS (1983) *Cleaners' Pay.* Study 290, May. London: Incomes Data Services

Jackson S (1970) Women in poverty. *The Times* 2 September

Kelly J (1988) *Trade Unions and Socialist Politics.* London: Verso

McCrindle J and Rowbotham S (eds) (1979) *Dutiful Daughters: Women Talk about their Lives.* Harmondsworth: Penguin Books

Pai H-H (2004) The invisibles—migrant cleaners at Canary Wharf. *Feminist Review* 78(1):164–174

Paul J (1986) *Where There's Muck There's Money: A Report on Cleaning in London.* London: Industry and Employment Branch, Greater London Council

Pearson P (1985) *Twilight Robbery: Trade Unions and Low Paid Workers*. London: Pluto Press

Pym J (ed) (2000) *Time Out Film Guide*. London: Penguin Books

Rowbotham S (1999) *A Century of Women*. London: Penguin Books

Rowbotham S and Beynon H (2001) *Looking at Class: Film, Television and the Working Class*. London: Rivers Oram

Rowbotham S, Segal L and Wainwright H (1979) *Beyond the Fragments: Feminism and the Making of Socialism*. London: Merlin Press

The London to Edinburgh Weekend Return Group (1980) *In and Against the State*. London: Pluto Press

Walker M (1972) Office cleaners face an uphill struggle. *The Guardian* 18 August

Chapter 12

The Privatization of Health Care Cleaning Services in Southwestern British Columbia, Canada: Union Responses to Unprecedented Government Actions

Marcy Cohen

Introduction

Some forms of privatization are gradual and incremental. This, however, was not the case with the privatization of housekeeping services in health care facilities in southwestern British Columbia (BC), Canada. In less than 10 months—from October 2003 to July 2004—all housekeeping services in the 32 hospital and extended care facilities in southwestern BC were contracted out to three of the largest multi-national service corporations in the world—Compass, Sodexho, and Aramark.[1] The impact on wages and working conditions was immediate and stunning: wages for the privatized housekeepers were cut almost in half, benefits were eliminated or drastically reduced, and union protections abolished (Cohen and Cohen 2004a). Overnight, these workers went from being the highest-paid to the lowest-paid housekeepers working in health care in Canada, such that their new rates are between 14% and 39% lower than anywhere else in Canada, and 26% below the national average (Table 1). The size and scope of this privatization places BC at the international forefront in the privatization of health support housekeeping services, outpacing even the United States and Britain, countries which have had over two decades of experience in privatizing such sectors—in the US, for example, in 2003 only 6.8% of hospitals reported that they contracted housekeeping services (Towne and Hoppszallern 2003:56), down from 27.4% in 1999 (Sunseri 1999:48), whereas in Britain, where competitive tendering for support services has been mandatory for close to two decades, private contractors controlled only 30% of the market in 2003 (Unison, Bargaining Support 2003:1).[2]

Table 1: Inter-provincial wage comparison of hospital and long-term care housekeeper (union rates) and Aramark/IWA rates, 1 April 2003

	IWA/Aramark Van Coastal Health Authority	BC's health support subsector	AB	Sask	Man	Ont	Que	NB	NS	PEI	Nfld	National average (union) wage rate 2003
Cleaner (C$)	10.25	18.32	13.60	13.22	12.74	16.82	14.29	12.73	11.92	13.40	12.28	13.93
% more than IWA/Aramark		44	24.6	22.5	19.5	39	28.2	19.5	14.0	23.5	16.5	26.4

Wage rate source

British Columbia (BC)   Health Services & Support Facilities Subsector Collective Agreement
Alberta (AB)   Alberta wage derived from average of CUPE Multi-Employer Agreement
Saskatchewan (Sask)   Sask wage derived from average of CUPE Sask and SEIU Saskatoon
Manitoba (Man)   Manitoba wage derived from weighted average for seven CUPE hospitals
Ontario (Ont)   Average rates of Ontario CUPE (OCHU) and Independents (info source SALAD, CUPE Research)
Quebec (Que)   CUPE (Quebec Federation of Labour)
New Brunswick (NB)   Average wage of the NB/CUPE Hospital Agreement, expired 30 June 2003, and NB/CUPE Nursing Homes Agreement
Nova Scotia (NS)   CUPE rates: clerical, service, and health care agreements
Prince Edward Island (PEI)   Average of CUPE Master, IUOE 942 Master and PEI Public Sector
Newfoundland (Nfld)   CUPE/NAPE and hospital boards

This privatization of housekeeping services was part of an untested and massive social experiment initiated by a newly elected provincial Liberal government determined to reinvigorate the private sector and to turn back what they viewed as the "unfair" gains made by public sector health care unions (Lavoie 2002:A4). The government claimed that health care agreements signed by public sector unions during the 10 years that the social democrats (ie the New Democratic Party [NDP]) were in power were negotiated through backroom "sweetheart deals". The legislation the Liberals introduced to facilitate privatization unilaterally altered the existing collective agreements of BC's health care unions. This legislation has no precedent in the history of labour relations in Canada, and has been subjected to Charter Challenge in the Supreme Court of Canada.

To make sense of how and why the provincial government in BC chose such a bold and, I would argue, reckless strategy, it is important to understand something about the history and character of the Hospital Employees' Union (HEU). The HEU has for the last 60 years represented health support workers in BC. The pay equity gains made by this union, their partisan support for public health care, and their determination to hold fast to their militant and solidaristic traditions are all aspects of this still unfolding story. On the other side of the ledger, there was a newly elected provincial Liberal government determined to foster private sector delivery of publicly funded services, multi-national service corporations who were more than willing to provide health support services using a low-paid contingent workforce, and a local branch of the private sector union—the Industrial, Wood and Allied Workers of Canada (IWA)—that was anxious to gain access to a new sector to compensate for its dwindling membership in the forestry sector (Greenwood 2003). In explaining these events, it is important to understand both how the groundwork for the acceptance of the privatization of health support services was laid prior to the election of the provincial Liberal government in 2001 and then how, once elected, the government orchestrated these massive changes in union rights at the political, administrative, and legislative levels.

The story, however, does not end here. Once the work was privatized, the struggle moved to a new level as the HEU became very active in organizing the new workforce and in monitoring the cleaning standards of the private contractors. And, despite the considerable efforts of the provincial government, the local government health authorities, and the private contractors to ensure that the workers would be represented by a private sector union, the largely female and new-immigrant workforce chose, at the vast majority of worksites, to re-organize with the HEU. However, the harsher working conditions, lower levels of compensation, poor training, and heavier workloads endured by the privatized workers militate against the provision of a quality service (Stinson, Pollak and Cohen 2005). This creates significant new challenges for both the HEU and government, as the union attempts to bargain its first collective agreements with the private contractors. The government, in turn, faces growing concerns from the public and from health care professionals concerning deteriorating cleaning standards in hospitals and long-term care facilities.

The paper is organized historically, tracing the social pre-conditions to privatization, the legislative and administrative processes through which privatization was achieved, and the post-privatization challenges for both labour and government. The goal of the paper is to provide the reader with a sense of both the political possibilities for, and limitation on, union activism in an era where neoliberal politics and economics are dominant.

Prelude to Privatization: The Hospital Employees' Union and its History

The Hospital Employees' Union is the largest health care union in BC, representing all hospital and the majority of long-term residential care workers employed in non-professional health support, direct care, and technical classifications (eg care aides, licensed practical nurses, clerical staff, trades and technical workers, housekeepers, and food service workers). Eighty-five percent of these workers are women, many of whom are the primary wage earners for their families (McIntyre and Mustel 2002). A high proportion of these women are older, visible minority and/or from immigrant backgrounds. The union has a long history of social unionism and hence of linking union issues to broader social and public policy. Over the years, HEU has worked with coalitions in building opposition to privatized health care and has been very successful in turning bargaining on privatization "into a political debate over public policy" (Johnson 1994:12). In the 1970s, for instance, the HEU launched a very successful organizing drive in the private nursing homes sector, a drive which included a very public campaign linking the poor working conditions of staff to the poor caring conditions for residents in private nursing homes, and which called for increased government regulation and not-for-profit delivery (Webb 1994).

The HEU has also been a leader in bargaining for pay equity for its membership. Historically, women working in the health care sector in BC earned significantly less than men doing similar work or work of equivalent value. HEU's struggle to redress the gender wage gap has spanned several decades and has been remarkably successful (Cohen and Cohen 2004b:21). The first inroad was made in the 1960s when the HEU succeeded in getting discriminatory "male" and "female" job classifications eliminated for people doing identical work (Webb 1994). During the 1970s, the union pursued several different strategies to achieve pay equity, including bargaining, filing human rights complaints, lobbying, and pursuing arbitrations. It was during this period that separate rates for male cleaners and female housekeepers were eliminated through a monthly anti-discrimination adjustment for the more than 8000 hospital workers earning less than the male cleaner rate.[3] Nevertheless, despite these gains, in 1991 there was still a wage gap between men and women cleaners of 10% and 29%, depending upon the specific job description (Fairey 2002). Consequently, in 1992, shortly after the election of the social democratic NDP government in BC, the HEU undertook a major strike where the key issue was pay equity. As a result of that strike, 90% of the HEU's membership received pay equity increases in addition to general wage increases. Although this did not establish full pay equity, it did mark the beginning of a process that was gradually improved upon throughout the 1990s. By 2001, the wage gap was less than 4% in most classifications, and the wage differential between different job families (ie between trades, technical, direct care work, clerical, and health support work) within the HEU had narrowed considerably (Fairey 2002).

Two things are worth noting here with regard to the HEU's campaign (Fairey 2003). First, the collective bargaining strategy of the HEU was much more effective in achieving pay equity gains for hospital support workers in BC than was the strategy pursued in Ontario, namely reliance upon government-implemented pay equity legislation. The overall average improvement for hospital support workers in BC was almost five times greater than that in Ontario. Second, and even more significant, is the fact that whilst pay equity adjustments greatly reduced the differential between the low- and high-wage earners in health support occupations in BC, in Ontario they actually increased these differences (ie women at the top of the wage scale received larger pay equity settlements than did women at the bottom of the wage scale).

The Right Responds

In 1995 the Fraser Institute, a well-known Canadian right-wing think tank headquartered in Vancouver, BC, published a slim, five-page "study" comparing the costs of ancillary support services in hospitals—cleaning, laundry, food services, trades, and clerical—to "hospitality" services in hotels. On the basis of this comparison, they concluded that hospital support workers in BC were overpaid (Ramsey 1995). In distinguishing "core public services" from "non-core domestic services" as a basis for arguing that non-core services should be "market tested", the Fraser Institute was articulating a strategy for privatization which had been pioneered by the Thatcher government in 1980s Britain (Jaffey 2005). Significantly, in the run-up to the 2001 provincial election these arguments were picked up and elaborated upon by a number of high-profile people in the media and by the BC Medical Association, who argued that the high wages paid to the "non-professional" and "non-essential" health support "hospitality" workforce were starving the province's health system of resources that should more rightly go to direct patient care and to health care professionals (Palmer 2000a, 2000b).

To address the designation of health support workers as merely "hospitality workers", the HEU commissioned Marjorie Cohen, a well-known labor economist and chair of the Women's Studies Department at Simon Fraser University, to prepare a report critiquing the Fraser Institute study by asking the question "Do Comparisons Between Hospital Support Workers and Hospitality Workers Make Sense?" (Cohen 2001). In order to determine if the actual work performed by hospital support workers is comparable to the work done in hotels, Cohen conducted interviews and focus group with more than 60 hospital and long-term care support staff in the fall of 2000. Based on the findings from these interviews and focus groups, she argued that health care support staff work is, in fact, significantly different from hotel work in that it requires "specialized skills" and training that is very specific to the health care environment. These skills are most often acquired on the job through a long-term attachment in the hospital sector. With regard specifically to housekeeping, Cohen argued that the work of health care housekeeping cannot be easily compared to the work performed by housekeeping staff in hotels, for the standards of cleanliness, the contact with patients, the complexity and technical sophistication of the physical environment, and the health risks for housekeeping staff are substantially different than in hotels (Cohen 2001:7–9). Based upon her own research, Cohen (2001:i) concluded that the Fraser Institute assessment was derived from "faulty research, heroic assumptions, and extrapolations that exaggerate the wage differentials between these two sets of workers". In contrast to the Institute, she argued (Cohen 2001:9) both that the higher wages for health care housekeepers were indeed justified and that the long-term stability and skills of the workforce were a direct consequence of the wages, benefits, pensions, and job security provisions negotiated by the unions in the health care industry, such that any reduction in these provisions would likely have a deleterious effect upon the quality of health care available in the province.

Whilst Cohen focused upon differences in skills between health care and hospitality workers, others have challenged the assumptions that privatization always leads to tax-payer savings. Sclar (2000:13), in his review of several cases of the privatization of "blue collar" services in the US over the last 100 years—and particularly in the last 20 years—has argued that "because of the complex nature of most public services, privatization contracts are typically written for multi-year periods [and that therefore], they foreclose easy competitive access to alternative providers if the product is not up to par". Specifically, he has documented some of the problems related to service quality (eg the quality of workmanship and willingness to respond to changing circumstances) that typically arise in multi-year contracts and has pointed out that the costs of contracting generally increase over time—whilst in the early years of privatization there may appear to be cost savings, over the longer haul the cost of privatized services is often greater than maintaining such services as publicly funded activities. He concludes, then, that "there is no easy, market-tested method for ensuring that citizens" get the services they want in a cost-effective manner (Sclar 2000:157).

Perhaps not surprisingly, however, the difficulties of market testing and comparisons with the private sector raised by academic economists such as Cohen and Sclar were not covered in the media. Instead, what did receive additional media attention were updated versions of the Fraser Institute Study, together with an inter-provincial comparison of wages of health care support workers prepared by the Health Employers Association of British Columbia (HEABC) (Government of British Columbia 2001; Ramsey 2002). According to a HEABC inter-provincial wage comparison, wages in BC were from 20% to 40% higher than for the rest of Canada. Cohen acknowledged that the wages for health support workers were, in fact, higher in BC than in other jurisdictions but argued that this situation was the result of the fact that overall wage rates in BC are higher than in the rest of Canada and that the wages for health support workers were merely in line with the province's higher costs of living, rather than being "disproportionately" higher (Cohen 2003). To back up these arguments she compared wages and costs of living for housekeepers in Toronto and Vancouver in 2002, showing that whilst a hospital housekeeper in BC was paid almost 11% more than a hospital cleaner in Ontario, housing costs were more than 12% higher in Vancouver than they were in Toronto, and the minimum wage was 16.8% higher in BC (Cohen 2003). Put another way, hospital housekeepers were actually underpaid in BC, relative to the overall cost of living. Again, these arguments received only minimal attention in the mainstream media.

To place this discussion within the wider political context, it is important to acknowledge that the promotion of public–private partnerships and the outright privatization of public services was an explicit policy objective of the newly elected Liberal government.

These policy directions reflect the commonly held assumptions of neoliberal governments internationally over the last 25 years, namely that public services (eg social, educational and health services) are best delivered by private, as opposed to public, entities, and that the private sector is always in a better position to control labor costs whilst still maintaining quality of services. Implementation of such a policy of privatization would be one of the first initiatives undertaken when, in 2001, after nine years of social democratic governance by the NDP, a coalition of politicians from the right and center was elected under a Liberal banner (winning 77 out of 79 seats in the provincial legislature), a victory achieved in large measure because of a success-ful media strategy to discredit the "state-building" activities of the previous NDP government. The goal of reducing government expen-ditures by privatizing services was quickly initiated through enactment of a series of legislative proposals which resulted in a C$2 billion tax cut that favored high income earners and a commitment to balance the budget over a three-year period. Significantly, this same impera-tive would not apply to the funding of physicians' services. Indeed, despite their public pronouncements that it wished to cut government expenditures, the Liberals advocated providing sufficient additional funding to ensure that physicians' compensation rates would be made equal to, if not higher than, the rest of the country (Ministry of Health Services 2004). Thus, in a household flyer produced three years after the election outlining steps it was planning to take to improve health care, the government would proudly announce that compensation for BC doctors had increased on average C$50,000 since the Liberals had taken over (Ministry of Health Services 2004). Significantly, this is considerably more than the yearly full-time wage and benefit package for the average HEU health support worker. The justification for the stark differentiation in treatment, then, appears to reflect the neoliberal view that market forces should govern compensation rates in health care. In the Liberals' view, whereas health professionals were essential to the health care system and in short supply, health support staff were simply unskilled hospitality workers in ready supply, a view graphically expressed by Kevin Krueger, the Liberal MLA (Member of the Legislative Assembly) from Kamloops, who has referred to hospital housekeepers as merely "toilet bowl cleaners" (Beatty 2004:A5).

Making Privatization a Reality: The Role of Government, Multinationals, and Unions

The mainstream media coverage, combined with the support garnered by the Fraser Institute "study", set the stage for the provincial govern-ment to introduce legislation to facilitate the privatization of health support services. In January 2002, nine months after they were elected into office with their huge plurality, the government passed Bill 29. This legislation unilaterally altered existing collective agreements negotiated between employers and unions representing approximately 100,000 workers in the health and social service sectors by removing the negotiated employment security protections and contracting-out protections in these agreements, as well as substituting language that allowed employers to lay off the existing workforce with minimum notice and avoiding union rights to follow the work (Government of British Columbia 2002). Such a change essentially left the employer free to restructure the workplace with an entirely new workforce paid at much lower rates and with far fewer benefits.[4]

The impact of this legislative change was huge. Thus, although in the previous 20 years there had been a significant increase in both federal and provincial government legislation limiting public sector bargaining rights through the use of increasingly restrictive back-to-work legislation (Fudge and Brewin 2005:28–29), and whilst back-to-work legislation for public sector workers has a long history in Canada, legislation aimed at altering collective agreement provisions has been exceedingly rare and, where it has occurred, it has usually been limited to changes in compensation rates (Rose 2003:15). Indeed, in an affidavit submitted with regard to Bill 29, Joseph Rose, a professor in the Faculty of Business at McMaster University, noted only three other occasions in Canadian history where governments infringed on statutory or collectively bargained job security provisions. In all of these cases government interventions were intended "to limit or foreclose" future bargaining on job security, but they did not "void collective agreement provisions during their term" (Rose 2003:17). In this respect, then, the provisions of Bill 29 were unprecedented. In response to the Bill's passage and its disproportionate impact on older, immigrant women workers in occupations like housekeeping, public sector unions representing health workers launched a court challenge, arguing that the Bill violated the Canadian Charter of Rights provisions on three grounds: equality rights (Section 15), freedom of association (Section 2), and security of persons (Section 7). This challenge was turned down at the BC Supreme Court in September 2003, but was heard by the Supreme Court of Canada in February 2006.

In addition to implementing legislation designed to undercut the unions' abilities to bargain collectively, the government prepared the ground for privatization in a number of other ways. Prior to introducing Bill 29 they dramatically reduced the number of regional health structures, thereby creating more centralized administrative structures and the economies of scale that would be attractive to large-scale external vendors. To further facilitate the contracting out of support services the province refused to fund the second and third year of the collective agreements which had been negotiated by health care workers just prior to the defeat of the NDP. Instead, health authorities were told to realign their priorities to compensate for the funding shortfall. Indeed, one of the very specific requirements outlined in performance contracts with health authorities was the requirement that by 2004–5 the administrative and support costs (eg for housekeeping, food and laundry services) should be 7% below what they were in 2001–2.

With the passage of Bill 29 and requirement to reduce costs in the health support sector, health authorities—primarily in the Lower Mainland and Southern Vancouver Island—began to put in place formal processes that would facilitate the contracting out of laundry, housekeeping, food services, and security services. The HEU, in turn, encouraged its membership to oppose privatization and to take action, up to and including civil disobedience. In addition, the union sought support from the broader labor movement and organized a very public media campaign to expose the poor record of service of the multi-national corporations who provided housekeeping, food, and laundry services for health care institutions in other jurisdictions (Hospital Employees' Union 2003).[5] In May 2002, taped evidence showed that one of the largest multi-national service corporations, Sodexho, was shopping around for a compliant union to circumvent the possibility that the HEU would organize these members (Hospital Employees' Union 2002a). When this became known, the overwhelming majority of the affiliates to the BC Federation of Labour, other than one local of the Industrial, Wood and Allied Workers of Canada (IWA), backed the HEU's right to organize this work, refused to co-operate with the outside contractors, and came out publicly in opposition to such backroom deals.

Throughout 2002, the HEU maintained its opposition to concessionary bargaining. Although the union took a somewhat more conciliatory position in the not-for-profit long-term care sector, in the larger and more significant hospital sector the HEU leadership chose an overtly militant course, including civil disobedience. Hence, in response to the first major privatization of laundry services by one of the Hospital authorities in October 2002, the HEU leadership decided to block privatized laundry trucks as a means of symbolically protesting the shipping of hospital laundry from the lower mainland of BC to a private company located in Alberta (Hospital Employees' Union 2002b). Three of the HEU's senior political leaders were arrested and charged, though they were later released. Then, on 27 January 2003, one year after Bill 29 was passed, the HEU organized a half-day of protest, picketing hospitals across the lower mainland and asking other union leaders to join them in a political protest against Bill 29. Although relatively successful, these protests were hardly enough to reverse the drive to privatize HEU work. In the aftermath of the protest, the HEU called for province-wide discussions with the employers to avert further contracting out. In March, the HEABC agreed to meet with HEU and, on 17 April 2003, a "tentative framework agreement" was negotiated between the HEU leadership and the HEABC, an agreement which put some tentative limitations on contracting out. Having negotiated some respite on the issue of contracting out, the HEU leadership argued that some wage concessions would now be needed to limit further contracting out. However, for the majority of the HEU membership that were not directly threatened by contracting out, the sudden shift to concessionary bargaining did not seem justified. As a result, when the time came to ratify the agreement, it was rejected by 57% of those who voted. For the licensed practical nurses, trades and other technical staff whose skills were in short supply, the idea of concessionary bargaining was anathema. The very short time-line, the lack of input on the contents of the package, and uncertainties related to the guarantees that contracting out would be capped at 3700 FTEs (full-time equivalents), made many others skeptical. In short, the solidaristic strategy that had been so successful during the previous 30 years did not hold under these conditions. Moreover, the expectations of the leadership—that the concessions should be shared more or less equally by all—created new schisms within the HEU membership.

Following the rejection of the concession package, the health authorities in southwest BC moved quickly to contract out support services with the industry's three largest multi-national private service corporations—Sodexho, Compass, and Aramark. Significantly, these corporations all operate internationally (having head offices in France, Britain, and the US, respectively), and all have reputations for poor labor relations and/or union bashing (Walker 2002a, 2002b, 2002c). By July 2004, approximately 8500 of the HEU and the BCGEU's housekeepers, food service, laundry, and security workers had lost their jobs (approximately 4000 of these workers were housekeepers and 95% were members of the HEU). With the expiration of the HEU's master collective agreement on 31 March 2004, the union was now in a position to strike legally.[6] During the negotiations the employer put concessionary demands on the table that were equal to or greater than the concessions already rejected by the HEU's membership in the previous year. The HEU, in turn, called for—and received—a strong strike mandate from its membership.

Having made their position clear, on 25 April the union's members across the province walked off the job. In the early hours of 29 April, the BC legislature passed Bill 37 which forcibly imposed a collective agreement on the union that included approximately 15% in hourly wage concessions, concessions which were very similar to those rejected a year earlier. The union refused to return to work and continued picketing in what was now an illegal strike. Over the next two days, grassroots union members across the province—primarily from public sector unions, but also from locals of private sector unions—walked out in support of the HEU workers. Public sympathy for the union appeared to be growing daily, with more and more unions planning to take their members out on the following Monday. In the end, four days into the illegal work stoppage, a deal was brokered by the BC Federation of Labour to limit further contracting-out to 600 jobs (although it did not reverse the contracting out already slated for June 2004) and to provide severance pay to those workers who would lose their jobs after 3 May 2004. It did nothing, however, to alter the approximately 15% wage cut contained in Bill 37. When the HEU members learned of the deal, many of them felt betrayed by the union and the leadership of the labor movement. This was particularly the case for the members who worked in trades, technical or direct care (ie licensed practical nurses), whose skills were in short supply and who therefore did not personally feel threatened by priva-tization. The divisions this created within HEU's membership, and in particular the concerns raised related to the lack of membership input into the decision, has spurred a review of the internal structures of the union, as well as changes in the senior leadership. Furthermore, whilst these events have not resulted in a shift away from the union's public commitment to social unionism and its open opposition to the privatization of health care services, the union does now face new challenges in negotiating for a much more diverse and divided workforce.

Organizing the Privatized Housekeepers, Monitoring the Contractors

In response to the growth in the amount of contracted-out health support work, the HEU has followed two sometimes contradictory strategies. On the one hand, they have mounted a very determined campaign to organize the newly privatized health support workers employed by the multi-national service providers and to gain support from the Labour Board in declaring the IWA agreements invalid. This organizing drive and decisions from the Labour Board have gone in HEU's favour, despite the obstacles put in its way by the govern-ment, the multi-national corporations, and one local branch of the IWA. At the same time, the union has continued with its very public research and communications strategy to expose the service quality problems created when health-related support services—and house-keeping in particular—are privatized.

HEU's Organizing Drive

The sub-standard wages and working conditions in the contracted-out jobs could not have been achieved without the co-operation of an "employer friendly" union. Prior to contracting out, the multi-national employers tried to establish a relationship with another union to fore-close the possibility that the HEU would mount a successful organizing campaign. As I have already noted, the majority of the unions affiliated to the BC Federation of Labour stood firm in opposition to the employers' attempts to circumvent the HEU's right to organize these workers. There was, however, one notable exception: Local 1–3567 of the IWA, a forestry union with a declining male membership and no previous history in representing women or health support workers. Local 1–3567 was quite willing to enter into "voluntary recognition agreements" with the multi-national employers (Aramark 2003). This development has been important because in such "voluntary recognition agreements" the terms and condition of employment are established by mutual agreement between the union and company before the new workforce is even in place, depriving the workers of a choice of trade union.[7] To this end, Local 1–3567 (IWA) signed unprecedented six-year voluntary recognition agreements with each of the three multi-national service corporations (by way of contrast, the vast majority of collective agreements in Canada are two to three years in length) (Aramark 2003).

The agreements reached between the employers and the IWA were all quite similar. In each case a "partnership agreement" established wage rates that the IWA has never before tolerated for its core, male membership—the median rate being around C$10.25 (Cohen and Cohen 2004b:18). In addition to the low wage rates, these agreements had no pension, long-term disability, job security or parental leave provisions, and vacation, sick time, and medical benefits were minimal (Aramark 2003). In addition, in each case people interested in applying for these positions were expected to attend job fairs, organized by the contractors, where they were interviewed and then required to attend an information session and sign a union card with Local 1–3567 (IWA) before being hired. As a result of these machinations, the workers had no opportunity to choose the union that would represent them or to vote on their collective agreement.

From the outset the HEU refused to recognize the validity of these "partnership agreements" and began an organizing drive as soon as the new workforce was in place. Organizers were hired from among the health support workers who had been laid off and who were representative of the ethnic composition of the new workforce. In tactics reminiscent of first contract organizing drives in other sectors, the union organized a member-to-member grassroots campaign. Organizers approached the new workers at or near the work site, talked to them about the poor wages, benefit provisions, and working conditions provided by the contractors and agreed to by Local 1–3567, and asked them to join the HEU. When the HEU had well over 45% of the workforce signed up (the percentage required by law to file for certification), they submitted these cards to the BC Labour Relations Board and filed for certification. In the period between late 2003 and early 2005, the HEU applied to represent housekeepers and food service workers in almost all of the hospitals and many of the long-term care sites where services were privatized. Each certification was opposed by both Local 1–3567 and the employers.

The HEU challenged the legality of each "voluntary recognition agreement" before the BC Labour Relations Board. On 20 May 2004, in the Board's first ruling on these matters, it agreed with the HEU and declared the voluntary recognition agreement between Local 1–3567 and Aramark housekeeping services in Vancouver Coastal Health Authority invalid. The reasons given by the Board were similar to those raised by HEU, namely that the workforce had no opportunity to choose its union or to vote on the provisions of the collective agreement (O'Brien, Leffler and Martin 2004). However, following this decision, the contractors continued to file complaints with the Labour Board against the HEU's right to represent these workers, thereby further delaying the vote count that would confirm or deny the HEU's right to represent these workers. In January 2005, the Board ordered that the first votes be counted, and since then the HEU has won applications to represent more than 3000 housekeeping and food services workers, with the support of some 75–100% of the workforce. By summer 2005, HEU was in a position to negotiate first collective agreements with both Aramark and Sodexho.

One of the significant outcomes for HEU in all of this is that, as a result of privatization, the union is now having to deal with a quite different labor force than before. A recent study has illustrated that the majority of these privatized workers are immigrant and visible minority women with family responsibilities and postsecondary education who, because of systemic discrimination in the Canadian labor market, are vulnerable to poverty wages and harsh working conditions provided by the contractors (Stinson, Pollak and Cohen 2005:42). In comparison to the previous in-house support staff, then, they are much more likely to be new immigrants with financial responsibilities for children and families overseas (67% as compared to 17%) and much more likely to have a second job (42% as compared to 9%). Due to unpredictable job assignments, routine understaffing, lack of training, and work overload, over 80% of the study participants reported that their physical health had been adversely affected by the job. Many talked about "feeling too rushed to work safely" and taking shortcuts "that put them at risk for needlestick and other occupational injuries" (Stinson, Pollak and Cohen 2005:7). Others reported that their supervisors were harsh, ill-informed, unsympathetic, and either unwilling or powerless to help with problemsolving, undoubtedly a result of the fact that many of the company supervisors lacked knowledge of the proper procedures required for cleaning areas with a high risk of infection and yet were prohibited by the contracting rules from working directly with in-house nursing and infection control staff (Stinson, Pollak and Cohen 2005:37). For the HEU, improving wages and establishing basic union rights that would allow these workers to address the issues of health and safety and appropriate training standards are critical first steps to addressing these issues. At the same time, learning how to relate to a new type of labor force presents the union with some significant challenges.

Monitoring the Private Contractors

Prior to the massive privatization in BC, Britain was the jurisdiction that had had the greatest experience with privatized health support services. There the public played a significant role in raising concerns about deteriorating standards of cleanliness and increased infection control risks resulting from the contracting out of hospital cleaning services (Department of Health [UK] 2001). Largely as a consequence of this public outcry, in 2001 the Labour government reversed the requirement for competitive tendering of support services that had been in force since 1983 (Department of Health [UK] 2001) and, in October 2004, it banned two-tier wage contracts in the National Health Service which allowed hospitals to buy cleaning services from multi-national companies at lower-than-union rates. This shift in policy reflected the growing recognition that it is difficult to maintain high cleaning standards with high turnover rates and low wages (Davies 2005).

Significantly, as soon as housekeeping services were contracted out in BC, a smattering of articles and letters began to appear in local newspapers with reports from staff and patients concerned about deteriorating standards of cleanliness in health care facilities. When these concerns were raised by members of both the BC Nurses Union and the HEU, the unions decided to jointly conduct a more systematic review of the cleaning standards, training, management, and monitoring processes that were put in place by the private contractors and the health authorities. To this end, the unions conducted a detailed survey of all staff in the emergency department at St Paul's Hospital in Vancouver, BC, interviewed previous HEU cleaning staff and members of the infection control team, and analyzed the monitoring processes put in place by the health authority (Pollak 2004). Based on this research, the union identified a number of systemic problems with the contracting process and called for an independent audit of all contracts. Around the same time, a woman who nearly died from an infection following a Caesarean section at Surrey Memorial Hospital went public with her concerns about the level of cleanliness at the hospital since contracting out (Canadian Broadcasting Corporation 2004). Following her disclosure, a number of other women came forward with similar concerns and the government was forced to order an investigation of the hospital. Shortly thereafter, the Vancouver Coastal Health Authority (VCHA) announced a three-week "external" audit of the housekeeping contracts in its region. The media and public interest in this audit was considerable. In just one day (17 December 2004) a joint news release from the HEU and the BCNU critiquing the VCHA "external" audit was quoted in 11 community newspapers, in the *Vancouver Sun*, and in a syndicated Canadian Press story.

Whilst the public profile on this issue has clearly shifted in favor of the HEU, the union nevertheless still faces considerable challenges. It must balance the often contradictory pressures of trying to establish enough trust with the contractors to bargain first collective agreements whilst at the same time continuing to expose the problems associated with privatizing hospital cleaning services. Although the union's ultimate goal is to bring services back in-house, the new members who are working for the contractors need the support of the union now to obtain a living wage and to gain even basic union rights.

Conclusion

The fact that a provincial government was able, in three short years, to create sweatshop-like conditions for thousands of workers in what was previously well-paid and secure public sector work is both disturbing and sobering. It points to the fundamental limitations in the legal and political structures of countries like Canada to protect its workforce against the worst aspects of global capitalism.

And yet, despite these very significant setbacks, HEU has not abandoned its commitment to equity and social unionism. The success of the grassroots organizing drive with immigrant and visible minority women and the willingness of the union to raise the public policy issues related to infection control and cleanliness are a testament to the union's ongoing commitment to equity and social unionism, and are indicative of the fact that these strategies can be quite effective even in the worst of times.

However, there are limits to what HEU can achieve on its own. The ongoing capacity of the union to successfully organize new members and to garner public opposition to contracting-out requires action that goes well beyond the workers and the union itself. To win further gains it will be critical for the union to join forces with other unions and community groups not only in BC and Canada but in other countries where these multi-national corporations operate.

Endnotes

[1] During this same 10-month time period, the health authorities also contracted out hospital food services (in the Vancouver Coastal Health Authority and in the Southern Vancouver Island Health Authority), and hospital security services (in the Fraser Health Authority and Vancouver Coastal Health Authority). In addition, over a somewhat longer time frame, support services (ie both housekeeping and food services) in many for-profit long-term care facilities were contracted out. These changes, like those in the hospital sector, occurred primarily in the lower mainland and in Southern Vancouver Island.

[2] It is also important to note there is no parallel for contracting out health care housekeeping services even in Canada. Calgary, where housekeeping services in two of four hospitals are privatized, is the next in size.

[3] The term "housekeeper" has since become the generic term to describe cleaners, since 80% of these workers are women.

[4] Although the legislation covers all of health care, it has primarily been used against the unions representing non-professional direct care and support staff and not the unions representing professional nurses or paramedical professionals.

[5] Public opposition to the contracting-out of support services was 55%, with 37% in support of contracting-out. This was up from 41% in June 2002 (Hospital Employees' Union 2003).

[6] Under its terms, the union was not allowed to engage in a strike during the period of operation of the master agreement.

[7] Voluntary recognition agreements are most common in the building trades and in forestry work, where work is short-term and specific trade unions have long-established records in protecting workers' rights in these industries. In these limited cases, setting up a "voluntary recognition agreement" between the employer and the trade union protects workers from having to build a union from the beginning each time a new short-term job begins. Instead, it guarantees them the wages and benefits already standard in the sector. This is a very different circumstance from the work in hospitals, where voluntary recognition agreements are undercutting wages in an established sector where an ongoing work relationship with a different union already exists (Cohen and Cohen 2004b:18).

References

Aramark Canada Facilities Services Ltd and Industrial Wood and Allied Workers of Canada, Local 1–3567 (2003) Collective agreement. http://www.lrb.bc.ca/cas/WQA15.pdf (last accessed April 2005)
Beatty J (2004) Liberal MLA nicknamed "Foghorn" has knack of choosing wrong words. *Times Colonist* 5 May:A5
Canadian Broadcasting Corporation (2004) Housekeeping problems at Surrey Memorial Hospital. CBC News, 19 November. http://www.vancouver.cbc.ca/regional/servlet/View?filename=bc_surrey-memorial20041119 (last accessed April 2005)
Cohen M G (2001) *Do Comparisons between Hospital Support Workers and Hospitality Workers Make Sense?* Vancouver, BC: Hospital Employees' Union
Cohen M G (2003) *Destroying Pay Equity: The Effect of Privatizing Health Care in BC*. Vancouver, BC: Hospital Employees' Union

Cohen M and Cohen M G (2004a) The politics of pay equity in the BC health care system: The role of government, multinational corporations and unions. *Canadian Woman Studies* 23 Spring/Summer:72–76

Cohen M G and Cohen M (2004b) *A Return to Wage Discrimination: Pay Equity Losses Through the Privatization of Health Care.* Vancouver, BC: Canadian Centre for Policy Alternatives—BC Office

Davies S (2005) Hospital contract cleaning and infection control. Commissioned by UNISON. http://www.unison.org.uk/acrobat/14564.pdf (last accessed November 2005)

Department of Health (UK) (2001) *National Standards of Cleanliness for the NHS.* London: Department of Health

Fairey D B (2002) "HEU's wage equity bargaining history." Unpublished paper commissioned by Hospital Employees' Union

Fairey D B (2003) An Inter-provincial Comparison of Pay Equity Strategies and Results Involving Hospital Services and Support Workers. Vancouver: Trade Union Research Bureau

Fudge D and Brewin J (2005) *Collective Bargaining in Canada: Human Right or Canadian Illusion?* Ottawa and Rexdale, Ontario: National Union of Public and General Employees and United Food and Commercial Workers Canada

Government of British Columbia (2001) Health care spending can't be exempt from cuts: The issue is not how much is spent; it's how well the system can treat British Columbians when we're sick. *Vancouver Sun* 22 September:A22

Government of British Columbia (2002) The Health and Social Service Delivery Improvement Act. Victoria, BC: Queen's Printer. http://www.qp.gov.bc.ca/statreg/stat/H/02002_01.htm (last accessed April 2005)

Greenwood, J (2003) Forest union fighting to remain a player. *Financial Post* 6 October

Hospital Employees' Union (2002a) *Blacklist Tapes also Reveal Evidence of Tendering Improprieties, HEU Charges.* Vancouver, BC: Hospital Employees' Union

Hospital Employees' Union (2002b) *HEU Leadership Arrested at Chillawak Laundry Blockage.* Vancouver, BC: Hospital Employees' Union

Hospital Employees' Union (2003) *Opposition Growing to Contracting Out.* Vancouver, BC: Hospital Employees' Union

Jaffey M (2005) Interview with Margie Jaffey, UNISON Researcher on Privatization of Public Services (unpublished). London, UNISON, 11 August

Johnson P (1994) *Success While Others Fail: Social Movement Unionism and the Public Workplace.* New York: ILR Press

Lavoie J (2002) Negotiator denies Liberal claims of NDP sweetheart deal. *Times Colonist* 5 February:A4

McIntyre and Mustel (2002) *HEU Member Profile Survey.* Vancouver, BC: McIntyre & Mustel Research Ltd

Ministry of Health Services (2004) Health care today and tomorrow. http://www.healthservices.gov.bc.ca/bchealthcare (last accessed April 2005)

O'Brien J, Leffler M and Martin G (2004) Formal decision BCLRB No. B173/2004. Vancouver, BC: British Columbia Labour Relations Board. http://www.lrb.bc.ca/decisions/B173$2004.pdf (last accessed April 2004)

Palmer V (2000a) Redistribute funds and privatize non-essential services says BCMA President. *Sounding Board* 39(8). http://www.boardoftrade.com/vbot_sb_archive.asp?pageID=179&IssueID=47&ArticleID=285&year=2000&sbPage=AD (last accessed April 2005)

Palmer V (2000b) NDP policies depleted the health system. *Vancouver Sun* 4 November:A22

Pollak N (2004) *Falling Standards, Rising Risks: Issues in Hospital Cleanliness and Contracting Out.* Vancouver, BC: British Columbia Nurses' Union and the Hospital Employees' Union

Ramsey C (1995) *Labour in the Hospital Sector.* Vancouver, BC: The Fraser Institute

Ramsey C (2002) *Labour Costs in the Hospital Sector.* Vancouver, BC: The Fraser Institute

Rose J (2003) "Affidavits in The Supreme Court of British Columbia." Unpublished documents in reference to *The Health Services and Support-Facilities Subsector Bargaining Association et al v Her Majesty the Queen et al*

Sclar E (2000) *You Don't Always Get What You Pay For: The Economics of Privatization.* New York: Cornell Univeristy Press

Stinson J, Pollak N and Cohen M (2005) *The Pains of Privatization: How Contracting Out Hurts Support Workers, Their Families and Health Care.* Vancouver, BC: Canadian Centre for Policy Alternatives—BC Office

Sunseri R (1999) Outsourcing on the outs. *Hospitals and Health Networks* 73(10):46

Towne J and Hoppszallern S (2003) 13th Annual contract management survey. *Hospitals and Health Networks* 77(10):52

Unison, Bargaining Support (2003) NHS market structure. http://www.unison.org.uk/acrobat/B828.pdf (last accessed April 2005)

Walker T (2002a) *Who is Sedexho?* Vancouver, BC: Hospital Employees' Union

Walker T (2002b) *Profile of Aramark Worldwide Corporation.* Vancouver, BC: Hospital Employees' Union

Walker T (2002c) *Profile of the Compass Group, PLC.* Vancouver, BC: Hospital Employees' Union

Webb P G (1994) *The Heart of Health Care: The First 50 Years.* Vancouver, BC: Hospital Employees' Union

Chapter 13

Justice for Janitors: Scales of Organizing and Representing Workers

Lydia Savage

Introduction

It is now commonly accepted that the labor movement in the US is in crisis, with only 8% of private sector workers and 12.9% of all workers in unions. As a result, in recent years much attention has been given to the development of new models of organizing workers. Typically, the new models embraced by labor leaders—at least rhetorically—as representing the labor movement's salvation are "bottom-up" in nature and call for a high level of participation by rank-and-file workers. Within this discussion of how to revitalize the labor movement, one model which has gained almost mythic status is that of the Justice for Janitors (JfJ) campaign, which is presented as a successful example of bottom-up organizing that has rendered an often-unnoticed group of workers visible, has built and diversified union membership, and has improved working conditions and pay for janitors. For many labor activists, part of the JfJ model's appeal is the fact that it has been developed in the context of the expanding service sector rather than the contracting manufacturing sector. However, although the JfJ has been lionized for its success, its potential use as a template for other campaigns has also raised a number of issues, especially concerning where the locus of power should lie within the union movement. This has been particularly so because the union which developed it—the Service Employees' International Union—has recently argued that many of the unions which make up the American Federation of Labor-Congress of Industrial Organizations (AFL-CIO) should merge, thereby bringing about a concentration of power at the national level so as to counter the power of nationally and globally organized corporations, a strategy which seems at odds with the locally sensitive organizing characteristic of the JfJ.

This paper, then, explores some of the debates around these questions of the JfJ model and the geography of power in the US labor movement. I begin with a discussion of some of the challenges faced when organizing service sector workers. I then outline a number of issues which relate to the deeply geographically-scaled nature of unions and what this means for conflicts over organizing strategies and the exercise of political power within unions. This is followed by an examination of some of the scalar tensions which emerged in the JfJ's Los Angeles campaign, arguably the SEIU's most successful. Finally, I conclude with a discussion of how some of the same issues which emerged in Los Angeles have also manifested themselves as the SEIU has attempted to export its organizing model to the broader US labor movement.

Challenges to Union Organizing in the New Economy

During the past decade, the AFL-CIO and a number of its constituent unions have placed a high priority on organizing low-wage, service sector workers in the belief that focusing on the service sector will not only increase union membership but, given that this sector employs primarily women and workers of color, will also diversify it. Leaders have spoken of the move to broaden organized labor's membership base as a necessary condition for revitalizing labor unions and so of creating a labor movement that can address a broad range of economic and social concerns. Responses to this new emphasis on organizing have varied widely: some labor activists have embraced the opportunity to redesign strategies and organizing models and to change the philosophy that underlies their organizing efforts, whereas others have continued to use a more traditional model of organizing, one which has its origins in the labor upheavals of the 1930s. Many have adopted a hybrid approach, continuing to use the traditional model of organizing yet adding a few different tactics that are seen as representative of new organizing methods.

In order to understand the import of these new strategies and models it is first necessary to outline those they are intended to replace. In particular, it is vital to recognize that the organizing model that has typically been used for the past six decades emerged from the manufacturing and mining sectors—labor's mid-century stronghold—and that such a model was perfected within the context of the "business unionism" which developed in the early post-war period as unions built large bureaucracies geared towards providing members with services (Hecksher 1988). Understanding this history is significant for two reasons. First, the spatialities of manufacturing and mining workplaces are quite different from those of most service sector workplaces, and this has implications for the model's appropriateness in these latter workplaces. Second, whereas the "community unionism" of the 1930s saw rank-and-file members often deeply engaged in their local union's activities and decision-making, under the business unionism model policy decisions are chiefly made at the national level, thus rank-and-file participation is largely denied (Moody 1988). Although business unionism has certainly been quite successful in securing real improvements in workers' standards of living, the shift away from community unionism (a shift highlighted by the 1950 contract between the United Auto Workers and General Motors, a contract known informally as the "Treaty of Detroit") meant that unions did trade a reliance on worker activism that was deeply rooted in communities for a reliance on an organizational structure that has often weakened the involvement and commitment of the membership, a weakening whose consequences are now coming home to roost (Faue 1991; Heckscher 1988, 2000; Moody 1988).

The "traditional" model of organizing developed in the mid-twentieth century, then, has several characteristics (Green and Tilly 1987). Most specifically, it generally involves unions using paid organizers to target "hot shops" and to appeal to workers by emphasizing "bread and butter" issues (wages and benefits). Given their expense, organizing campaigns are invariably run in a technical, top-down fashion, with the emphasis being on quickly gaining the 51% of the vote necessary to win a certification election. Finally, organizers typically target large workplaces in industries where workers are thought to be easily unionizable, since this will result in more members in exchange for the union's organizing efforts. However, with the faltering of this model as the economy has shifted away from manufacturing, as the workforce has become more diverse, and as the geography of workers and workplaces has changed, pressures have grown for new models which challenge the tactics and, often, the spatial assumptions of the traditional model. These new models hold that, for the labor movement to be truly revitalized, unions must rethink their philosophy of organizing and building unions. This is frequently discussed as moving from "business unionism" or "service unionism" to an "organizing model of unionism", one in which organizing is continuous and high levels of rank-and-file worker participation, activism, and leadership are developed and maintained. It is, perhaps, no surprise that the low-wage service sector has been the source of many new tactics for organizing, the result of the fact that, by and large, the model developed in manufacturing and mining has not worked well for service workers (Bronfenbrenner et al 1998; Milkman and Voss 2004).

Several differences between manufacturing and mining and service sector workplaces, then, can be distinguished which have implications for the transferability of models of organizing from one economic sector to the other (Berman 1998; Gray 2004; Savage 1998). Primarily, service workplaces are typically smaller and more decentralized than are manufacturing/mining facilities. In addition, service workers are often scattered throughout many types of workplaces—for instance, janitorial staff may move through several buildings each night, while health care workers frequently go from home to home. As a consequence, service sector organizers must design tactics that will allow them to reach both, say, 2000 clerical workers employed in a single office park but also 20 janitors cleaning several different buildings in an urban downtown (Savage 1998). In seeking to develop such tactics, perhaps the most obvious challenge posed by a fragmented service workplace is the identification and contacting of workers, especially because the traditional methods of leafleting and approaching workers at factory entrance gates during shift changes is a tactic that is rarely successfully—unlike in manufacturing, service workplaces such as office complexes, universities, retail settings, and hospitals generally have multiple entrances and parking lots, service workers such as clerical staff or retail workers have varied shifts and often take their breaks according to the pace of the work, and many workers often adopt a style of dress indistinguishable from that of others in the workplace (in contrast to the coveralls of factory workers), such that workers targeted by organizers become part of the crowd of management personnel, customers, clients, students, or patients when entering and leaving the workplace. Equally, the fact that many service sector workers work at night in scattered worksites and remain invisible to most people also makes mass leafleting problematic.

In addition to such issues, how workplaces themselves are internally laid out can dramatically influence the relationships between workers and thus the possibilities for unionization. In particular, service sector workers often work more closely with their bosses than is the case in manufacturing or mining. For example, in her analysis of the spatial strategies of resistance employed by the clerical and technical workers union at Yale University, Lee Lucas Berman (1998) shows how the university administration consciously separated clerical and technical workers from each other physically, such that even within small shared offices workers were isolated by being placed in cubicles. Where isolation was not possible, workers were kept in open office spaces under full view of supervisors. As Berman illustrates, the result of such control of the micro-geography of the workplace was a social fragmentation of the workforce based upon its spatial fragmentation. Certainly, similar kinds of control may be attempted in the non-service sector. However, unlike in a small office or retail space, in a mine or manufacturing facility it is usually the case that workers have greater opportunity to find places beyond the gaze of their supervisors, places where they might "talk union" or simply goof off.

While the shift to service sector work has had an important impact upon labor organizing, neoliberalism and sweeping economic changes have also altered the employer-employee relationship in ways not accounted for in the traditional model. Hence, no longer do many service sector workers such as janitors work year round, full time for an employer directly. Rather, it is now commonplace to work as a temporary, part-time or contingent worker for an employer or a sub-contractor. Furthermore, the complex network of contracting relationships and the fact that corporations have become huge and diversified entities, operating in many different sectors of the economy with multiple types of employment relations, means it is now more important than ever to have good research about patterns of ownership and lines of corporate control. The dismantling of the vertically integrated corporation which dominated the economy in the 1950s or 1960s has, in other words, made it harder to determine where, ultimately, corporate decision-making power resides and upon whom workers must bring pressure to bear if they are to be successful. Thus, if any organizing campaign and union representation of workers is to be effective it must often spend significantly more time and money engaging in corporate research than was the case in the past.

Despite the many challenges posed by a restructuring economy, unions are crafting new models and strategies. The question in all this, however, is "will workers respond?" If early responses are any indication, the answer seems to be a resounding "yes". Thus, in their extensive study of union election outcomes, Bronfenbrenner and Juravich (1998) have shown that, of all factors, it is union tactics that play the most important role in explaining union election results, with unions' choice of tactics having a much greater impact on organizing outcomes than do anti-union efforts by employers or labor laws perceived by unions as unfavorable. Specifically, Bronfenbrenner and Juravich argue that the tactics that were most effective in winning elections were those which encouraged high rates of worker participation in the organizing campaign through housecalls, frequent meetings (large and small), organizer attendance at worker social functions, including workers in strategy design, and forming committees to work on bargaining issues before there is even an election—that is to say, tactics which run counter to the model which has dominated for the past half-century in which unions rely upon professional organizers. They caution, however, that housecalls are not a "magic bullet" and that these tactics are only effective when used within a campaign that emphasizes rank-and-file worker participation and involves the wider community, so that issues faced by workers at one worksite are not seen as isolated concerns but, rather, as concerns of the larger community. For sure, some commentators have argued that such "community unionism" is nothing new but is instead simply a long-overdue return to the roots of the labor movement (Cobble 1991; Wial 1993, 1994). Regardless of its historical origins, however, many unions have increasingly embraced community unionism as a model of representing workers at a time when the labor movement is grappling with dwindling membership. More importantly, such unions appear to be having some success (Banks 1991; Fine 2000; Johns and Vural 2000; Johnston 1994; Savage 2004; Tufts 1998; Walsh 2000; Wills 2001). These successes, however, have raised important questions concerning matters of union structure—for instance, should the power to devise strategies rest at the local level so that organizers can develop locally sensitive campaigns, or does it need to be co-ordinated at a national level so as to be able to match the organizational structure of employers who are increasingly national and/or international in scope, and what kinds of intra-union tensions do such questions spawn?

Institutional Scales and Worker Activism

While geographers have created a vibrant literature that examines the relationship between the spatial scales at which unions operate and how they affect or resist the economic and political policies articulated by firms and/or the state (Berman 1998; Castree et al 2004; Gray 2004; Herod 1991, 1997, 1998, 2001a, 2001b; Holmes 2004; Sadler 2004; Savage 1998, 2004; Savage and Wills 2004; Tufts 1998, 2004; Walsh 2000; Wills 1996, 1998a, 1998b; Wilton and Cranford 2002), they have generally been less concerned to examine unions as institutions which themselves have internal scales of power, authority, and decision-making. Thus the fact that there exist national union bodies (for historical reasons these are called "International unions" in the US/Canadian context), central labor councils, and local branches of International unions ("Locals") which frequently have quite different agendas points both to the inherently geographical nature of political organization and to the fact that there are frequently significant tensions between the various scales at which the labor movement operates. Furthermore, as the labor movement experiments with new forms of organizing, questions are emerging about what is the appropriate scale—local, regional, national, international—for decision-making and the exercise of power. In what ways and at what scales should the labor movement and its individual unions operate to be effective defenders of workers' interests yet also remain responsive to such workers? At what scales do they need to structure themselves in order to face the enormous challenges posed by an ever-changing global economy? How big can a union structure grow before worker activism and participation are no longer developed or supported?

These questions go to the heart of trying to avoid the emergence of a new bureaucracy which might stifle nascent efforts—such as those of the JfJ campaign—designed to reinvigorate the labor movement by breaking out of the dominant (and bureaucratic) business unionism model. This is particularly important because much research on social movements has argued that institutions which start out by challenging orthodoxy invariably succumb to what the German sociologist Robert Michels (2001/1915) labeled the "iron law of oligarchy"—once any social movement establishes a bureaucracy, there is a move away from the very radicalism that led to the creation of the institution and toward an interest to protect the *status quo*. However, Voss and Sherman (2000) have argued that this argument does not appear to fit recent moves to revitalize the US labor movement. Like many researchers, they identify a shift over the past two decades in some quarters from the pursuit of "business unionism" to the development of "comprehensive campaigns" which are characterized by a high level of activism by members and can eventually, though not inevitably, lead to an organizing model of unionism. Thus, in their study of Northern California unions, they found that, under certain circumstances, local unions can break free of the bureaucratic conservatism which has developed in the labor movement. Three factors appear to be most significant in avoiding bureaucratic tendencies. First, they found that unions that successfully challenged the iron law had experienced a political crisis of some sort which had led to a change of leadership through local elections or International union intervention. Second, these new leaders had experience in other types or forms of activism and the leaders took the decline of unionism as an opportunity—indeed, a mandate—to change strategies and to innovate. Finally, International—that is to say, national-level—unions facilitated this new activist leadership by providing the resources and support necessary for local leaders to succeed in implementing change.

The involvement of the International union, however, can have significant impacts upon the degree of freedom which union Locals may enjoy when it comes to developing innovative and locally sensitive organizing strategies. Certainly, even under the business unionism model most Locals in the US have long retained a measure of autonomy regarding such matters as the number of officers they have and the time, place, and frequency of their meetings. Furthermore, they generally have had responsibility for local collective bargaining and grievance procedures, as well as for choosing in which organizing campaigns they would engage and what tactics and strategies they would use.[1] But it is also important to remember that these decisions are made within the context of a Local's budget and resources, and Locals have long argued that they need more money and staff to run innovative organizing campaigns. Theoretically, since they pay a percentage of their dues to the International union, the International will return some of these monies in the form of personnel, subsidies, and training, and the support of one's International union is often critical in any innovative organizing campaign, both in terms of providing financial support but also in terms of granting legitimacy to a new leadership and/or institutional change. However, while their support is often needed to put the heft of the national-level union behind a local union in any particular dispute, such support usually comes at a price—national leaders frequently want to exert some degree of control over a Local's activities and are often criticized for engaging in what is seen as undue and "top-down" interference in the day-to-day operations of local unions. Consequently, much intra-union conflict revolves around struggles between the national and local leaderships over who should exercise ultimate power on particular issues, and many activists' efforts to secure the institutional spaces for engaging in innovative campaigns have focused upon expanding local union and rank-and-file control within the existing multiple scales of union decision-making concerning the availability of funding and resources, together with localizing the exercise of power within the overall union structure—the Teamsters for a Democratic Union campaign, for instance, represents one ongoing national effort to demand more rank-and-file participation in the International's day-to-day operations.

Though developing new models of organizing seems like a logical goal for unions if they are to reinvigorate the labor movement, such tensions between local unions, regional labor councils, International unions, and even the broader AFL-CIO has meant that the issues of what proportion of membership dues should be used to fund the servicing of current members relative to the amount which should be spent upon organizing new members, and at what level (local or national) should that decision be made, have become significant bones of contention. As a way to examine these issues, then, I now turn to one of the most widely documented comprehensive union organizing campaigns—the Justice for Janitors (JfJ) campaign (see Clawson 2003; Cranford 1998, 2004; Erickson et al 2002; Fisk, Mitchell and Erickson 2000; Howley 1990; Hurd and Rouse 1989; Lerner 1996; Mines and Avina 1992; Rudy 2004; Savage 1998; Waldinger et al 1998; Wial 1993, 1994; Wilton and Cranford 2002). Developed by the Service Employees' International Union (SEIU) in the mid-1980s, the JfJ campaign has been heralded for its success in a time of challenge for the US labor movement. The campaign, however, is not only important in its own right but it also has much broader implications. Specifically, the SEIU national leadership's frustration with what it saw as the failure of many of the leaders of the AFL-CIO's constituent

Internationals to organize new workers, together with the apparent unwillingness of the Federation's own leadership to force the issue, led it to propose in December 2004 its "Unite to Win" plan which would have fundamentally changed the way in which the broader labor movement operated by, amongst other things, giving the AFL-CIO authority to require co-coordinated bargaining by unions, to force mergers in sectors where several small unions represent workers, and to funnel more money to unions which engage in the kinds of innovative organizing represented by the JfJ campaign.

The SEIU's plan has, to say the least, been controversial—for some, such a concentration of authority was the only strategy which would allow unions to confront the power of corporations, whereas for others it represented nothing but a naked powerplay by the SEIU national leadership. In what follows, then, I explore some of the tensions between these contradictory desires for the centralization and decentralization of authority as they emerged in the JfJ's Los Angeles campaign. I then consider what these mean for the introduction of new organizing models such as the JfJ within the broader labor movement.

Justice for Janitors: A Comprehensive National Campaign

Unions have long represented building service janitors. In the 1980s, however, unions representing janitors (such as the SEIU) were faced with declining memberships as the demographics of the workforce changed and as the nature of the employee–employer relationship was transformed by the growth of sub-contracting. While some locals held onto their base memberships, many saw memberships plummet and it was clear that if the SEIU continued to organize and represent workers in the same way as it always had then the union would end up losing the industry completely to non-union contractors. The recognition of a need for change by the SEIU resulted in the JfJ campaign, a campaign hailed as one of the most innovative and comprehensive campaigns designed by an International union and one which has been quite successful in gaining new union members. Significantly, the SEIU has committed 30% of its resources to organizing and has taken the lead in national campaigns to organize, particularly, janitors and healthcare workers, campaigns that are characterized by disruptive tactics and militant public actions. Along with the increased financial commitment to organizing, the International has pushed local unions to organize new members and has, in fact, allocated resources based upon local organizing efforts. Often, SEIU organizers are dispatched from Washington, DC, to assist local unions and/or to direct their efforts. The end result is that the International has directed local unions to engage in more worker organizing, has subsidized many of these efforts, and has provided skilled staff and training to carry them out. Combining an innovative approach to organizing workers by geographical area rather than by worksite with well-publicized public actions and a commitment to representing immigrants (even if they are not legal immigrants), the JfJ campaign has sought to remove wages from competition—long a central goal of union organizers—by ensuring that all janitors in a defined area or district are unionized, so that the pressure on sub-contractors to underbid each other to win contracts with building owners is removed.

One of the earliest and most successful JfJ campaigns involves that of Local 399 in Los Angeles. Historically, the Local had primarily represented health care workers who worked mainly for Kaiser, a large Health Maintenance Organization (HMO). However, in 1987 the International encouraged it to begin organizing janitors instead. The result was a dramatic unionization of janitors in the city: whereas only 10% of LA's downtown commercial building service janitors were unionized at the start of the campaign, by 1995, 90% were. In addition to dramatically increasing union coverage amongst janitors, the Los Angeles campaign has been notable for the fact that many of the janitors are immigrants, most of them are women, and almost all are Latina/o—all groups which have traditionally been viewed by unions as difficult to organize. Union organizers made deep connections in the immigrant community and gathered strength for their actions by involving community groups, immigrant rights groups, and the personal networks that already existed among workers. The sensitivity paid to the identities of workers and the comprehensive campaign combined to result in the tremendous success of the JfJ campaign. The campaign's success, however, unleashed significant divisions within Local 399, divisions which are important to understand if the JfJ model is to be fully evaluated for its appropriateness for more widespread application to organizing campaigns.

Institutional Change and LA's "Reformistas"

As a result of JfJ's success, by the mid-1990s Local 399 had grown to represent 28,000 workers. However, for almost two decades, beginning in the mid-1970s, the Local's leadership had remained virtually unchanged. The rapid growth in membership, though, unleashed tensions within the union. Specifically, many of the long-time members who were healthcare workers wanted the union's focus to return to healthcare and servicing members, instead of having the union continue its efforts to emphasize the janitorial campaign. Additionally, numerous rank-and-file members from all occupations disliked the fact that many of the JfJ organizers came from outside the Local. Others, particularly new members, complained that the recent influx of Latinos meant that the leadership was not now representative of its members. Within this context, in June 1995 a 21-member slate calling itself the Multiracial Alliance/"*Las Reformistas*" was elected to the Local's executive board. Eleven percent of the Local membership voted in the elections and the slate won all 21 of the races contested on the 25-seat board. While the slate captured board seats up to the executive vice-president, they nevertheless faced opposition. The longtime president of Local 399, Jim Zellers, soon squared off with the new, largely Latino leadership. The new board passed a series of proposals, including the establishment of a grievance committee and the firing of some union employees. It also attempted to fire 12 of the Local's 80 employees and to hire new staff, though Zellers refused to comply, arguing that he alone had the right to hire and fire union personnel and that the new board was trying to usurp his authority.

Eventually, this internal dissension began to affect the Local's day-to-day running. Thus, whereas Zellers argued that things had run smoothly until the new executive board had arrived, the new board suggested that any conflict was the result of Zellers's refusal to abide by the outcome of the election and that, Solomon-like, if Zellers and the "old guard" truly cared about the Local, they would relent. After several weeks of such internal battles, in August 1995 the dissident slate began a hunger strike. Zellers and his supporters claimed the tactics of the dissidents were unacceptable, although ironically the union had supported these tactics when they had been used against an employer. Significantly, though, the hunger strike began at precisely the same time that SEIU President John Sweeney was running his own dissident campaign for the presidency of the AFL-CIO. As a way of gaining some control over the situation, on 14 September 1995 Sweeney suspended the newly elected officers and placed Local 399 in trusteeship. After suspending the *Reformistas*, he appointed Mike Garcia to serve as Local 399 trustee, for up to 18 months. Significantly, Garcia himself was head of SEIU Local 1877, a union headquartered more than 300 miles away which represented janitors in Oakland, Sacramento, and Silicon Valley, and which was itself also involved in the International's JfJ campaign, having been created when Locals 18 and 77 merged.

Sweeney's rationale for establishing a trusteeship was that the battle between the self-proclaimed dissidents and Local 399's president and his supporters was negatively affecting the union's ability to represent its members. In particular, the image of a union embroiled in internal turmoil was seen as both weakening the bargaining position of workers who were facing contract negotiations, and as tarnishing the image of a campaign that relied heavily on public support for its public and militant actions. This was exacerbated by the fact that the public was kept abreast of the news via the LA press, which was not kind to either the union or the dissident slate in its coverage of the internecine quarrel—reporters, for instance, frequently trivialized the issues by characterizing the Local as "crippled by a nasty spat between the president and his supporters and rival dissidents" (del Olmo 1995) and referring to the dissident slate as "rabble rousers" (Nazario 1995). Equally, whereas some union members intimated that the lead dissident, Cesar Sanchez, could not speak English well enough to negotiate contracts and lacked experience to do the job, others—such as hunger striker Martin Berrera—were quoted as saying that the old guard "treat us like ignorant peasants" (Nazario 1995). Perhaps most ominously, still others suggested that building owners were taking advantage of the disarray to try to go non-union.

Upon accepting John Sweeney's mandate, Garcia stepped in from his Northern California base and set up a new leadership in Local 399, replacing Zellers. He quickly became involved in the furor, making one of his few public statements about events in response to a column in the *LA Times* by Associate Editor Frank del Olmo (1995), who had asked "if the Latinos and women who were unionized by the JfJ campaign are not yet ready to assume leadership, the question for the union is, when will they be?" Significantly, del Olmo had concluded his column with the opinion that innovative campaigns are often painful for local unions but that this is a predictable outcome of the "growth and change that is inevitable once formerly all-male, all-Anglo institutions open their doors to large numbers of minorities and women. After all, once you help raise the consciousness of workers for the first time, it is naïve to assume that they will use their newfound skills only to criticize their employers"—the implication being that union leaders might also expect criticism for their apparent high-handedness. In response, Garcia (1995) defended his trusteeship by stating that "Local 399 had ceased to function because of the dispute". Pointing out that the leaders of the Latino insurgents had been offered full-time jobs in the union and "an opportunity to develop their skills and leadership abilities during the rebuilding of the local union, after which self-governance and democracy will be restored", Garcia argued that his appointment, along with that of a team of deputy trustees, to oversee the Local was, at the time, the only possible solution to the crisis.

Significantly, this was not the first time that Garcia had been handpicked for an assignment by the International union. He had become an organizer for the SEIU in 1980. In 1985, when the SEIU designed the JfJ strategy, the International's leadership had selected him to be the point man for their efforts, largely because he had already been working with janitors and was one of the few SEIU staff members who spoke Spanish. Given this background, some Local 399 members felt he was more tied to the interests of both the International and the JfJ campaign than the concerns of the Local's membership. For many, this fear was realized when, within his first year as trustee, Garcia split Local 399 into two unions, removing the janitors from Local 399 and putting them into his own, Local 1877. As a result, Local 399's *Reformistas* lost active and committed dues-paying members, and Local 1877 gained them. As a result, questions concerning how active workers could be in leadership and daily operations in a Local union headquartered 300 miles away from their place of work immediately rose to the fore. In response, Garcia argued that "every time [Local 399] tried to organize in healthcare, the janitorial side fell down and every time you tried to organize janitors, the healthcare side fell down".[2] The answer to this problem, Garcia suggested, was to have a single, statewide Local focused on one industry. This would allow the Local to match the scale of organization of the industry, an industry in which there is increasing consolidation among building services and property services corporations. Thus, for Garcia:

> It makes sense as much as practically and as reasonably possible to adjust [union] structures . . . It [the increased size of the Local] creates a lot more power to leverage companies for the benefit of workers and working families. Size gives you raw power and using your leverage with [building] owners, clients, renters, leasers, and sub-contractors makes a difference in organizing . . . It all comes down to leverage in different areas at different levels. There's political leverage, community leverage, legal leverage, and industry leverage. Aramark [an international company which provides food, beverage, and cleaning services for a range of educational, health-care, and other businesses] is our Wal-Mart, Compass [a food service company operating in more than ninety countries] is another and they are entering our traditional areas of cleaning. Aramark and Compass combined have as many workers as Wal-Mart. It takes a discipline of focus, resources, and strategic thinking [to challenge them].

In support of his argument, Garcia has contended that the combination of grassroots activism from mobilized members and statewide power has let Local 1877 move faster and be more powerful on recent organizing efforts. For example, he suggests that the success of JfJ in Los Angeles (including a strike in 2000) has given the union power and momentum to organize janitors in locations long thought too difficult to organize. Thus, drawing upon their LA triumphs, JfJ has successfully organized janitors in Sacramento and Silicon Valley and taken on corporations such as Hewlett-Packard. JfJ is also now in suburban San Diego, where, as Garcia point outs, "2,000 Orange County workers were organized in 2001 in a place most people thought you could never organize, given the political landscape".

Indeed, according to Garcia, Orange County and San Diego "show how things can move when you are large and powerful".

While it may or may not make sense to match corporations with statewide Locals, balancing the needs and wants of members with such union structures is not always easy. Hence, in the statewide Local 1877, recent dissent has once again resulted in structural change. In 2004, the Local's San Francisco office decertified from the SEIU and founded a new, independent union. However, this event, at least according to Garcia, is a rarity and the union has learned much from the decertification: whereas in Los Angeles the merger between the two Locals was effective because the union leadership spent a lot of time with the membership and had a timeline and process in place, such that, in the end, even opposition forces agreed to the merger, in San Francisco the union spent little time educating the members about the benefits of the proposed merger with Local 1877. As a result of the lessons learned from the San Francisco decertification, Local 1877 has made changes to its structure, and key offices throughout the statewide Local now have vice-presidents elected by the union membership. For his part, though, Garcia believes that union members care first about "strong unions making a difference in incomes, healthcare, respect and dignity" and only secondly about "politics and elected officials". Thus, for him, "the tricky part of [running] a large Local, a statewide Local—what I realized in San Francisco—was I can't be everywhere ... There is no way I can be deeply involved in a community and membership with 7–8 local offices". The goal of this reorganization, then, is to ensure that the rank-and-file has some control over the Local's agenda by directly electing officials to key offices yet also to secure the power that comes with a large structure. Indeed, Garcia and other SEIU leaders contend that it is only by matching the 21st-century corporation in size and power *and* by creating and maintaining rank-and-file activism that not only the SEIU but also the US labor movement can survive and grow. As Garcia has put it: "where Andy [Stern, President of SEIU] has taken our union and wants to take the labor movement is where the global economy has gone". This goal of taking the labor movement where global capitalism has gone has informed the JfJ campaign but, perhaps more significantly, it is also the underlying tenet of the SEIU's "Unite to Win" plan and the resulting "Change to Win" coalition, which split from the AFL-CIO in the summer of 2005.

JFJ as a Model for the Future of the US Labor Movement?

Mirroring the "up-scaling" of its organizational structure in California, so that a union "Local" was now to be defined as having a statewide structure, the SEIU has recently proposed a significant concentration of power within the AFL-CIO as being the only way to challenge nationally and globally organized firms. As a result, all of the struggles over scalar politics that are part of the SEIU's experience in California have now emerged in the debate over the future direction of the entire US labor movement. Specifically, in response to what it saw as the failure of the labor movement to address declining membership, in 2004 the SEIU initiated its up-scaling campaign, outlining its strategy in a widely distributed document entitled "Unite to win: A 21st century plan to build new strength for working people". Seeking to replicate the JfJ campaign throughout multiple industries, SEIU leaders proposed that US unions consolidate and then organize and represent workers along occupational lines as a way to match the organizational structures of 21st-century employers. At the heart of this strategy was the belief that being scattered among multiple small unions weakens workers in two ways: first, there are many small unions that cannot, even with good leadership, match the resources of employers; and second, workers that share an industrial sector, craft or market suffer from fragmented bargaining power.

As evidence, the report noted that 15 different unions represent transportation and construction union members, 13 unions have significant numbers of public employees, there are nine major unions in manufacturing, while health care union members are divided amongst more than 30 unions. Moreover, most of the forty AFL-CIO national unions have fewer than 100,000 members, while 15 unions represent 10 of the 13 million members in the AFL-CIO. For the SEIU, then, a successful strategy for the future would be one in which the International unions of the AFL-CIO would "develop and implement a plan . . . to (1) unite the strength of workers who do the same type of work or are in the same industry, sector, or craft to take on their employers, and (2) insure that workers are in national unions that have the strength, resources, focus, and strategy to help nonunion workers in that union's primary area of strength to join and improve workers' pay, benefits, and working conditions". To achieve these goals, the SEIU proposed a significant centralization of power with the AFL-CIO,[3] such that its

> Executive Council should have the authority to recognize up to three lead national unions that have the membership, resources, focus, and strategy to win in a defined industry, craft, or employer, and should require that lead unions produce a plan to win for workers in their area of strength. In consultation with the affected workers, the AFL-CIO should have the authority to require coordinated bargaining and to merge or revoke union charters, transfer responsibilities to unions for whom that industry or craft is their primary area of strength, and prevent any merger that would further divide workers' strength. The

unions of the AFL-CIO should work together to raise pay and benefit standards in each industry. Where the members of a union have clearly established contract standards in an industry or market or with a particular employer, no other union should be permitted to sign contracts that undermine those standards.

The initial proposal by the SEIU was countered with plans suggested by other unions and affiliated groups, such as the Communications Workers of America, and the American Federation of State, County, and Municipal Employees. In contrast to the SEIU's plan, such unions have argued that significant changes must be made but unions should retain their autonomy and the AFL-CIO should not be granted such sweeping power. Many union activists, leaders, and rank-and-file members have also questioned the proposal to change the structural nature of the US labor movement without first changing its philosophical nature and breaking the hold that the idea of business unionism still has over many union officials and workers. Thus, in January 2005 the A. Philip Randolph Institute, the Asian Pacific American Labor Alliance, the Coalition of Black Trade Unionists, the Coalition of Labor Union Women, the Labor Council for Latin American Advancement, and Pride At Work issued a "Unity Statement" pointing out that, while organizing is recognized as the challenge:

> those responsible for organizing decisions and for leading organizing campaigns frequently do not include people of color and women. Also, the tremendous challenge to organize people of color in the South, in the Southwest, and in diverse urban areas lacks adequate support and resources. The labor movement should not assume that nonunion workers lack any organization. Indeed many workers of color and immigrant workers participate in their community through civic, religious, and other forms of "identity-based" organizations that are potential allies of the labor movement. Time and attention to cultivate labor and community alliances to support organizing are crucial. The constituency organizations are uniquely positioned to build strong, enduring bridges of solidarity between unions and civil rights, religious, women's, immigrant, minority and Lesbian, Gay, Bisexual and Transgender organizations.

Such tensions over the movement's future finally erupted in July 2005, when the SEIU, the International Brotherhood of Teamsters, the Laborers' International Union of North America, UNITE HERE, the United Food and Commercial Workers International Union, the United Brotherhood of Carpenters and Joiners of America, and the United Farm Workers—unions representing over 5 million workers— pulled out of the AFL-CIO to form a new federation, the "Change to Win Coalition". Such a development has raised many questions about the labor movement's future path. How can workers build a labor movement that is sensitive to the grassroots and encourages the participation in decision-making of the rank-and-file union membership yet which can also counter globally organized capital? In what ways should the leadership structure of the labor movement and individual unions change to reflect the new face of the laborforce? If the labor movement scales up its strategies and structures without simultaneously supporting local members and activists and their needs, does the labor movement run the risk of recreating business unionism on a larger scale? Significantly, many of these questions revolve around the issue of geography, specifically at what spatial scale should power reside, and how might unions develop strategies which will allow them both the greatest flexibility and give them the greatest power to confront the unevenly developed economic geography of global capitalism. Although virtually all observers agree that organizing and mobilizing workers is key to any effort to revitalize the movement and that significant resources must be directed to such efforts, the question remains how to achieve these goals and what must be sacrificed to do so. Hence, whereas many see it as ironic that a union hailed for a campaign such as the JfJ that was so innovative, radical, and locally specific would launch a plan that consolidates power in the hands of a few unions, others argue that the labor movement must match organizationally its employers, who are increasingly national and international in structure.

Conclusion

Decentralized workplaces that scatter workers across more and more worksites are becoming increasingly common, forms of employer–employee relations are becoming more varied, and the paid workforce in the US is more diverse today than it has been in a century. If it is to be successful in defending workers' rights, then, the labor movement needs to create models of organizing that can address all of these issues. Moreover, research repeatedly points to the fact that organizing efforts succeed more often when there is high rank-and-file participation in the campaign. Yet, in order to motivate workers to participate in union activities, organizers must understand the workplace, the workers, and the employer. Thus, while campaign issues may be similar across sectors and place (intensification of the labor process, for instance), they take on particular forms in different workplaces. Likewise, the types of actions which are appropriate will vary according to the workers—for instance, the mass marches of red-t-shirted workers and supporters favored by LA janitors may not appeal to janitors in other places. The key, therefore, seems to be the ability to develop campaigns and structures which provide sufficient flexibility to incorporate local specificity, yet which also provide enough collective mass and centralized coordination as to allow workers to stand up to their employers.

The labor movement, therefore, clearly faces many decisions about organizing. One-on-one organizing is frequently very successful, but is an expensive and lengthy process. Equally, while local autonomy is often critical for organizing efforts, parent unions are usually unwilling to provide funding without retaining some degree of control—often, quite considerable—over the campaign. Creating a union or campaign characterized by high levels of worker participation means that paid union staff must relinquish some decision-making powers and let the membership set goals and policy, but this is a difficult task in a labor movement in which power is deeply entrenched. Furthermore, the diversity of the workforce poses challenges as union organizers and unions struggle with institutional racism and sexism, the most obvious evidence of which is that the present makeup of the US labor movement's leadership and staff is a far cry from reflecting the diversity of the service workforce. From elected positions on executive boards to paid positions as organizers, white men hold positions of power at higher rates than their membership in particular unions.

Models of organizing such as the JfJ campaign emerging from the service sector appear to point the way toward successfully increasing both membership and participation. It is critical that unions organize new workers and increase levels of worker participation and activism. What remains to be seen, though, is at what scale decisions will be made and how union culture and structures will change as a result of changing organizing strategies and a changing membership. In building the groundwork for a new labor movement, it is critical that the values of unions are shaped by workers in relation to their particular workplace culture, the employer, and the identities of workers. It is also critical that they carefully consider how to build the institutional structures that allow for sufficient collective action to challenge nationally and globally organized capital.

Acknowledgments

A version of this paper was presented at the Annual Meeting of the Association of American Geographers in New York in March of 2001. I would like to thank Jane Wills for co-organizing the sessions with me and Luis Aguiar and Andrew Herod for the opportunity to be a part of this collection. I am grateful to Luis, Andy, Melissa Gilbert and two anonymous reviewers for comments that greatly strengthened the paper. Mike Garcia and Jonno Shaffer made time to speak with me despite very busy schedules and I am very appreciative.

Endnotes

[1] Many unions have national "master" contracts which are negotiated by the International union and cover such matters as wages and length of vacation time but which allow Local unions to develop locally specific terms and conditions which modify the master contract on certain matters (such as the specific times at which workers may take breaks during a shift).

[2] Unless otherwise indicated, all quotations by Garcia are from interviews with the author, conducted on 5, 7 and 9 March 2005.

[3] This is a significant change, for the AFL-CIO as an institution has long had a decentralized structure, with power residing with the individual member unions rather than with the Federation itself. This contrasts with the model in countries such as Germany, where the central labor federation—in the case of Germany, the Deutscher Gewerkschaftsbund (DGB: German Confederation of Trade Unions)—has more power to tell individual member unions what to do.

References

Banks A (1991) The power and promise of community unionism. *Labor Research Review* 18:7–31

Berman L (1998) In your face, in your space: Spatial strategies in organizing clerical workers at Yale. In A Herod (ed) *Organizing the Landscape: Geographical Perspectives on Labor Unionism* (pp 203–224). Minneapolis: University of Minnesota Press

Bronfenbrenner K and Juravich T (1998) It takes more than housecalls: Oganizing to win with a comprehensive union-building strategy. In K Bronfenbrenner, T Juravich, S Friedman, R Hurd, R Oswald, and R Seeber (eds) *Organizing to Win: New Researchon Union Strategies* (pp 19–36). Ithaca, NY: ILR Press

Bronfenbrenner K, Juravich T, Friedman S, Hurd R, Oswald R and Seeber R (eds) (1998) *Organizing to Win: New Researchon Union Strategies.* Ithaca, NY: ILR Press

Castree N, Coe N, Ward K and Samers M (2004) *Spaces of Work: Global Capitalism and Geographies of Labour.* London: Sage Publications

Clawson D (2003) *The Next Upsurge: Labor and the New Social Movements.* Ithaca: ILR/Cornell University Press

Cobble D (1991) Organizing the postindustrial work force: Lessons from the history of Waitress Unionism. *Industrial and Labor Relations Review* 44(April):419–436

Cranford C (1998) Gender and citizenship in the restructuring of janitorial work in Los Angeles. *Gender Issues* 16(4):25–51

Cranford C (2004) Gendered resistance: Organizing Justice for Janitors in Los Angeles. In J Stanford and L F Vosko (eds) *Challenging the Market: The Struggle to Regulate Work and Income* (pp 309–329). Montreal and Kingston: McGill-Queen's University Press

del Olmo F (1995) Perspective on labor: Empowerment has its problems; the sudden prominence of the service workers' union belies the trouble the old guard has with its base of minorities and women. *Los Angeles Times* 29 October:5

Erickson C, Fisk C, Milkman R, Mitchell D and Wong, K (2002) Justice for Janitors in Los Angeles: Lessons from three rounds of negotiations. *British Journal of Industrial Relations* 40:543–567

Faue E (1991) *Community of Suffering and Struggle.* Chapel Hill: University of North Carolina Press

Fine J (2000) Community unionism in Baltimore and Stamford: Beyond the politics of particularism. *WorkingUSA* 4(3):59–85

Fisk C, Mitchell D and Erickson C (2000) Union representation of immigrant janitors in Southern California: Economic and legal challenges. In R Milkman (ed) *Organizing Immigrants: The Challenge for Unions in Contemporary California* (pp 199–224) Ithaca: Cornell University Press

Garcia M (1995) Service union trusteeship. *Los Angeles Times* 7 November:8

Gray M (2004) The social construction of the service sector: Institutional structures and labour market outcomes. *Geoforum* 35(1):23–34

Green J and Tilly C (1987) Service unionism: Directions for organizing. *Labor Law Journal* August:486–495

Heckscher C (1988) *The New Unionism: Employee Involvement in the Changing Corporation.* New York: Basic Books

Herod A (1991) Local political practice in response to a manufacturing plant closure: How geography complicates class analysis. *Antipode* 24:385–402

Herod A (1997) From a geography of labor to a labor geography. *Antipode* 29:1–31

Herod A (1998) *Organizing the Landscape: Geographical Perspectives on Labor Unionism.* Minneapolis: University of Minnesota Press

Herod A (2001a) Labor internationalism and the contradictions of globalization: Or, why the local is sometimes still important in the global economy. *Antipode* 33:407–426

Herod A (2001b) *Labor Geographies: Workers and the Landscapes of Capitalism.* New York: Guilford Press

Holmes J (2004) Re-scaling collective bargaining: Union responses to restructuring in the North American auto industry. *Geoforum* 35(1):9–22

Howley J (1990) Justice for janitors: The challenge of organizing in contract services. *Labor Research Review* 15:61–72

Hurd R and Rouse W (1989) Progressive union organizing: The SEIU Justice for Janitors campaign. *Review of Radical Political Economy* 21:70–75

Johns R and Vural L (2000) Class, geography and the consumerist turn: UNITE and the Stop Sweatshops Campaign. *Environment and Planning A* 32:1193–1213

Johnston P (1994) *Success While Others Fail: Social Movement Unionism and the Public Workplace.* Ithaca: ILR Press

Lerner S (1996) Reviving unions. *Boston Review.* April/May

Michels R (2001/1915) *Political Parties: A Sociological Study of the Oligarchical Tendencies of Modern Democracy.* Batoche Books: Kitchener, Ontario (first published in English in 1915)

Milkman R and Voss K (2004) *Rebuilding Labor: Organizing and Organizers in the New Union Movement.* Ithaca: ILR/Cornell University Press

Mines R and Avina J (1992) Immigrants and labor standards: The case of California janitors. In J Bustamante, C Reynolds and R Hinojosa Ojeda (eds) *US–Mexico Relations: Labor Market Independence* (pp 429–448). Stanford: Stanford University Press

Moody K (1988) *An Injury to All: The Decline of American Unionism.* London: Verso Press

Nazario S (1995) Hunger strike marks union's split labor: Strong dissident group within janitors organization launches protest after effort to take control of board is blocked. *Los Angeles Times* 8 August:1

Rudy P (2004) "Justice for Justice," not "compensation for custodians": The political context and organizing in San Jose and Sacramento. In R Milkman and K Voss (eds) *Rebuilding Labor: Organizing and Organizers in the New Union Movement* (pp 133–149). Ithaca: ILR/Cornell University Press

Sadler D (2004) Trade unions, coalitions and communities: Australia's Construction, Forestry, Mining and Energy Union and the international stakeholder campaign against Rio Tinto. *Geoforum* 35(1):35–46

Savage L (1998) Geographies of organizing: Justice for Justice in Los Angeles. In A Herod (ed) *Organizing the Landscape: Geographical Perspectives on Labor Unionism* (pp 225–252). Minneapolis: University of Minnesota Press

Savage L (2004) Public sector unions shaping hospital privatization: The creation of Boston Medical Center. *Environment and Planning A* 36(3):547–568

Savage L and Wills J (2004) New geographies of trade unionism. *Geoforum* 35(1):5–8

Tufts S (1998) Community unionism in Canada and labor's (re) organization of space. *Antipode* 30:227–250

Tufts S (2004) Building the "competitive city": Labour and Toronto's bid to host the Olympic Games. *Geoforum* 35(1):47–58

Voss K and Sherman R (2000) Breaking the iron law of oligarchy: Union revitalization in the American labor movement. *American Journal of Sociology* 106(2):303–349

Waldinger R, Erickson C, Milkman M, Mitchell D, Valenzuela A, Wong K and Zeitlan M (1998) Helots no more: A case study of the Justice for Janitors campaign in Los Angeles. In K Bronfenbrenner, T Juravich, S Friedman, R Hurd, R Oswald and R Seeber (eds) *Organizing to Win: New Research on Union Strategies* (pp 102–119). Ithaca: ILR/Cornell University Press

Walsh J (2000) Organizing the scale of labor regulation in the United States: Service sector activism in the city. *Environment and Planning A* 32:1593–1610

Wial H (1993) The emerging organizational structure of unionism in low-wage services. *Rutgers Law Review* 45:671–738

Wial H (1994) New bargaining structures for new forms of business organization. In S Friedman, R W Hurd, R A Oswald and R L Seeber (eds) *Restoring the Promise of American Labor Law* (pp 303–313). Ithaca: ILR/Cornell University Press

Wills J (1996) Geographies of trade unionism: Translating traditions across space and time. *Antipode* 28:352–378

Wills J (1998a) Taking on the cosmocorps: Experiments in transnational labor organization. *Economic Geography* 74:111–130

Wills J (1998b) Space, place and tradition in working-class organization. In A Herod (ed) *Organizing the Landscape: Geographical Perspectives on Labor Unionism* (pp 129–158). Minneapolis: University of Minnesota Press

Wills J (2001) Community unionism and trade union renewal in the UK: Moving beyond the fragments at last? *Transactions of the Institute of British Geographers* 26(4):465–483

Wilton R and Cranford C (2002) Toward an understanding of the spatiality of social movements: Labor organizing at a private university in Los Angeles. *Social Problems* 47(3):374–394

Notes on Contributors

Luis L M Aguiar is an associate professor of sociology at the University of British Columbia, Okanagan Campus. He researches, and has been doing so for the last decade, neoliberalism and its impact on immigrant and minority workers in the Canadian building-cleaning industry. In addition, he has written on whiteness, racism, and growing up immigrant in Montreal. At the moment, he is engaged in studying the Okanagan Valley and its changing status as a hinterland in the global economy. A research project on janitors' cross-border organizing is in development, as is a study of former Canadian boxing champion Eddie Melo, and pop diva Nelly Furtado. He teaches urban sociology, race, and ethnic relations in Canada, the sociology of tourism, cultural studies, qualitative methods, and globalization and labor.

Andries Bezuidenhout is a researcher in the Sociology of Work Unit at the University of the Witwatersrand, Johannesburg. His doctoral dissertation was on the workplace regimes in household appliance factories in South Africa, Swaziland, and Zimbabwe. He has published widely on areas such as labor market policy and industrial strategy. His current research interest is in the "postcolonial workplace regime", specifically how the processes of corporate restructuring in southern Africa are influenced by the drive to become globally competitive; the need to adhere to global labor, social, and environmental standards; and local pressure from the state and society to address the legacies of the colonial workplace regimes.

Anne Katrine Blangsted received her MSc degree in physical education/human physiology from the August Krogh Institute, University of Copenhagen, Denmark in 1998 and her PhD degree in muscle physiology in 2005 from the Faculty of Health Sciences, University of Copenhagen, Denmark and the National Institute of Occupational Health (NIOH), Denmark. She is currently employed as a researcher at the Department of Physiology at NIOH, Denmark. Her main research interest is human exercise physiology with a focus on work-related musculoskeletal pain, physical activity/inactivity, muscle activity, and muscle fatigue.

Alyson Brody lived and worked in Thailand for six years. She received her doctorate in social anthropology from the School of Oriental and African Studies in 2003. She is currently employed as a research officer at the Institute of Development Studies in the United Kingdom.

Rodrigo Hidalgo Dattwyler teaches in the Instituto de Geografía at the Pontificia Universidad Católica de Chile. He received his doctorate in Human Geography from the Universidad de Barcelona. His current research focuses on the production of urban residential spaces. He is the author of numerous publications on Santiago's urban landscape and is the editor of the *Revista de Geografía Norte Grande* and a member of the Editorial Advisory Board of *EURE*, a Chilean journal that deals with urban and regional studies.

Khayaat Fakier is a graduate researcher in the Sociology of Work Unit at the University of the Witwatersrand, Johannesburg. She is currently involved in a three-country study that focuses on the work and social reproduction of household appliance workers in South Africa, Korea, and Australia. Her research interests are in exploring how "locality" has affected the ways of understanding and acting of the female, working poor in distinct local worlds.

Lotte Finsen received her MSc degree in biology and physical education in 1991 and her PhD degree in physiology in 1995, from the University of Copenhagen. Until 2004 she has been working with basic and applied physiology at the department of Physiology at the National Institute of Occupational Health in Copenhagen. She is currently employed as a physiologist in the Occupational Health Service. Her research interests focus on mechanisms behind development of occupational-related musculoskeletal disorders, in particular within repetitive monotonous work.

Andrew Herod is Professor of Geography and Adjunct Professor of International Affairs and of Anthropology at the University of Georgia. He is author of *Labor Geographies: Workers and the Landscapes of Capitalism* (Guilford, 2001), editor of *Organizing the Landscape: Geographical Perspectives on Labor Unionism* (University of Minnesota Press, 1998), and co-editor of *Geographies of Power: Placing Scale* (Blackwell, 2002) and *An Unruly World? Globalization, Governance and Geography* (Routledge, 1998). He writes frequently on issues of labour and globalization. He was recently a Visiting Distinguished Professor at the University of Sydney in the Work and Organisational Studies Department.

Karen Messing, PhD, is an ergonomist, occupational health specialist, and Professor in the Department of Biological Sciences of the

Université du Québec in Montréal, Canada. Her current research focuses on applications of gender-sensitive analysis in occupational health, effects of prolonged standing, and constraints and demands of work in the service sector. She was co-founder and first director of CINBIOSE, a WHO-PAHO Collaborating Centre in Early Detection and Prevention of Work and Environment-Related Illness. Dr Messing co-directs a research partnership with three Québec unions oriented towards improvement of women's occupational health. She is the author of numerous articles and of *One-eyed Science: Occupational Health and Working Women* (Temple 1998) and editor of *Integrating Gender in Ergonomic Analysis* (1999), published in six languages by the Technical Bureau of the European Trade Union Confederation. She is on the editorial boards of the *International Journal of Health Services, Women and Health, Recherches féministes, Policy and Practice in Health and Safety*, and *Salud de los trabajadores*.

Sheila Rowbotham is Professor of Gender and Labour History in the Social Science School, Manchester University. She helped to found the women's liberation movement in Britain and has written extensively on women's and labour history.

Shaun Ryan is a lecturer in organisational behaviour and management at Curtin University and is completing his PhD in Work and Organisational Studies at the University of Sydney. His thesis examines employment relations and organisational culture in the commercial cleaning industry. He is currently working on a research project examining the work aspirations of Australian university graduates and the public sector as an employer of choice, as well as continuing his ongoing research into the organisation of work in the Australasian cleaning industry. Shaun has also published on union organisation strategies in the shearing industry in Australia and New Zealand and the gender revolution in New Zealand unions.

Lydia Savage is an associate professor of geography and chair of the Department of Geography-Anthropology at the University of Southern Maine where she is also a member of the Women's Studies Council and a founding member of the Labor Studies Minor Program. She earned her BA in geography from the University of California, Berkeley and her MA and PhD in geography from Clark University. Her current research examines the ways in which labor unions are reshaping union strategies and transforming institutional cultures in light of contemporary social, cultural, and economic change. A former member of the International Association of Machinists, she is currently a member of the Associated Faculties of the University of Maine System, MEA/NEA.

Ana María Seifert DESS (ergonomics), MSc, is an ergonomist, occupational health specialist, and researcher at CINBIOSE, Université du Québec à Montréal, a WHO-PAHO Collaborating Centre in Early Detection and Prevention of Work and Environment-Related Illness. She was a founding member of a research partnership with three Québec unions oriented towards improvement of women's occupational health. Ongoing studies include the analysis of space constraints on special education aides and of formal and informal strategies of health care workers for the prevention of infections in hospital settings. She is currently enrolled in a PhD programme in social and preventive medicine at Université Laval. She is the author of numerous articles on constraints and demands of work in the service sector. She is on the editorial board of *Salud de los trabajadores*.

Karen Søgaard received her MSc in physical education from the August Krogh Institute, University of Copenhagen and at the same institution she pursued a PhD in human physiology in 1994. She spent 8 months as a research fellow at the Department of Kinesiology at Simon Fraser University, Vancouver, Canada and four months at Prince of Wales Medical Research Institute in Sydney, Australia. Currently she is a Senior Researcher in the research group on Performance and Pain Physiology at the National Institute of Occupational Health, Denmark. Her research interest concerns the interactions between muscle pain and motor control in relation to work-related musculo-skeletal disorders.

Patricia Tomic is a Chilean Canadian who teaches sociology in the Irving K Barber School of Arts and Sciences at the University of British Columbia, Okanagan Campus. She earned her PhD at the Ontario Institute for Studies in Education, University of Toronto. Her areas of research, like herself, move between Canada and Chile. In Canada, she studies the politics of language and the Latin American immigrant experience, and also researches issues of "race" and representation in the hinterlands. In Chile, she studies neoliberalism and globalization by looking at higher education, women, and urban issues.

Ricardo Trumper is an Associate Professor of sociology in the Irving K Barber School of Arts and Sciences at the University of British Columbia, Okanagan Campus. He received his PhD in Social and Political Thought from York University. His present research includes globalization, neoliberalism, and post-fordism. He is currently working on mobility, fear, and sports in Chile and on the repercussions of neoliberalism on the Okanagan Valley in Canada.

Index

Note: Page numbers in italics refer to tables.